ELEMENTARY STOCHASTIC CALCULUS

with Finance in View

**ADVANCED SERIES ON STATISTICAL SCIENCE &
APPLIED PROBABILITY**

Editor: Ole E. Barndorff-Nielsen

Advanced Series on

Statistical Science &

Applied Probability

Vol. 6

ELEMENTARY STOCHASTIC CALCULUS

with Finance in View

Thomas Mikosch

Laboratory of Actuarial Mathematics
University of Copenhagen
Denmark

 **World Scientific**

NEW JERSEY • LONDON • SINGAPORE • BEIJING • SHANGHAI • HONG KONG • TAIPEI • CHENNAI

Published by

World Scientific Publishing Co. Pte. Ltd.

5 Toh Tuck Link, Singapore 596224

USA office: 27 Warren Street, Suite 401-402, Hackensack, NJ 07601

UK office: 57 Shelton Street, Covent Garden, London WC2H 9HE

Library of Congress Cataloging-in-Publication Data
Mikosch, Thomas.
    Elementary stochastic calculus with finance in view / Thomas
Mikosch.
        p.   cm. -- (Advanced series on statistical science & applied
probability ; v. 6)
    Includes bibliographical references and index.
    ISBN-13 978-981-02-3543-7
    ISBN-10 981-02-3543-7 (alk. paper)
    1. Stochastic analysis.   I. Title.   II. Series.
QA274.2.M54   1998
519.2--dc21   98-26351
                                                                    CIP

British Library Cataloguing-in-Publication Data
A catalogue record for this book is available from the British Library.

First published 1998
Reprinted 1999, 2000, 2003, 2004, 2006, 2008

Printed by FuIsland Offset Printing (S) Pte Ltd, Singapore

Preface

Ten years ago I would not have dared to write a book like this: a non-rigorous treatment of a mathematical theory. I admit that I would have been ashamed, and I am afraid that most of my colleagues in mathematics still think like this. However, my experience with students and practitioners convinced me that there is a strong demand for popular mathematics.

I started writing this book as lecture notes in 1992 when I prepared a course on stochastic calculus for the students of the Commerce Faculty at Victoria University Wellington (New Zealand). Since I had failed in giving tutorials on portfolio theory and investment analysis, I was expected to teach something I knew better. At that time, staff members of economics and mathematics departments already discussed the use of the Black and Scholes option pricing formula; courses on stochastic finance were offered at leading institutions such as ETH Zürich, Columbia and Stanford; and there was a general agreement that not only students and staff members of economics and mathematics departments, but also practitioners in financial institutions should know more about this new topic.

Soon I realized that there was not very much literature which could be used for teaching stochastic calculus at a rather elementary level. I am fully aware of the fact that a combination of "elementary" and "stochastic calculus" is a contradiction in itself. Stochastic calculus requires advanced mathematical techniques; this theory cannot be fully understood if one does not know about the basics of measure theory, functional analysis and the theory of stochastic processes. However, I strongly believe that an interested person who knows about elementary probability theory and who can handle the rules of integration and differentiation is able to understand the main ideas of stochastic calculus. This is supported by my experience which I gained in courses for economics, statistics and mathematics students at VUW Wellington and the Department of Mathematics in Groningen. I got the same impression as a lecturer of crash courses on stochastic calculus at the Summer School of the

Swiss Association of Actuaries in Lausanne 1994, the Workshop on Financial Mathematics in Groningen 1997 and at the University of Leuven in May 1998.

Various colleagues, friends and students had read my lecture notes and suggested that I extend them to a small book. Among those are Claudia Klüppelberg and Paul Embrechts, my coauthors from a book about extremal events, and David Vere-Jones, my former colleague at the Institute of Statistics and Operations Research in Wellington. Claudia also proposed to get in contact with Ole Barndorff-Nielsen who is the editor of the probability series of World Scientific. I am indebted to him for encouraging me throughout the long process of writing this book.

Many colleagues and students helped in proofreading parts of the book at various stages. In particular, I would like to thank Leigh Roberts from Wellington, Bojan Basrak and Diemer Salome from Groningen. Their criticism was very helpful. I am most grateful to Carole Proctor from Sussex University. She was a constant source of inspiration, both on stylistic and mathematical issues. I also take pleasure in thanking the Department of Mathematics at the University of Groningen, my colleagues and students for their much appreciated support.

Thomas Mikosch   Groningen, June 1, 1998

Contents

Reader Guidelines

This book grew out of lecture notes for a course on stochastic calculus for economics students. When I prepared the first lectures I realized that there was no adequate textbook treatment for non-mathematicians. On the other hand, there was and indeed is an increasing demand to learn about stochastic calculus, in particular in economics, insurance, finance, econometrics. The main reason for this interest originates from the fact that this mathematical theory is the basis for pricing financial derivatives such as options and futures. The fundamental idea of Black, Scholes and Merton from 1973 to use Itô stochastic calculus for pricing and hedging of derivative instruments has conquered the real world of finance; the *Black–Scholes formula* has been known to many people in mathematics and economics long before Merton and Scholes were awarded the Nobel prize for economics in 1997.

For whom is this book written?

In contrast to the increasing popularity of financial mathematics, its theoretical basis is by no means trivial. Who ever tried to read the first few pages of a book on stochastic calculus will certainly agree. Tools from measure theory and functional analysis are usually required.

In this book I have tried to keep the mathematical level low. The reader will not be burdened with measure theory, but it cannot be avoided altogether. Then we will have to rely on heuristic arguments, stressing the underlying ideas rather than technical details. Notions such as measurable function and measurable set are not introduced, and therefore the formulation and proof of various statements and results are necessarily incomplete or non-rigorous. This may sometimes discourage the mathematically oriented reader, but for those, excellent mathematical textbooks on stochastic calculus exist.

In discussions with economists and practitioners from banks and insurance companies I frequently listened to the argument: "Itô calculus can be understood *only* by mathematicians." It is the main objective of this book to overcome this superstition.

Every mathematical theory has its roots in real life. Therefore the notions of Itô integral, Itô lemma and stochastic differential equation can be explained to anybody who ever attended courses on elementary calculus and probability theory: physicists, chemists, biologists, actuaries, engineers, economists, ... In the course of this book the reader will learn about the basic rules of stochastic calculus. Finally, you will be able to solve some simple stochastic differential equations, to simulate these solutions on a computer and to understand the mathematical ideology behind the modern theory of option pricing.

> **What are the prerequisites for this book?**

You should be familiar with the rules of integration and differentiation. Ideally, you also know about differential equations, but it is not essential. You must know about elementary probability theory. Chapter 1 will help you to recall some facts about probability, expectation, distribution, etc., but this will not be a proper basis for the rest of the book. You would be advised to read one of the recommended books on probability theory, if this is new to you, before you attempt to read this book.

> **How should you read this book?**

It depends on your knowledge of probability theory. I recommend that you browse through the "boxes" of Chapter 1. If you know everything that is written there, you can start with Chapter 2 on Itô stochastic calculus and continue with Chapter 3 on stochastic differential equations.

You cannot proceed in this way if you are not familiar with the following basic notions: stochastic process, Brownian motion, conditional expectation and martingale. There is no doubt that you will struggle with the notion of conditional expectation, unless you have some background on measure theory. Conditional expectation is one of the key notions underlying stochastic integration.

The ideal reader can handle simulations on a computer. Computer graphs of Brownian motion and solutions to stochastic differential equations will help you to experience the theory. The theoretical tools for these simulations will be provided in Sections 1.3.3 and 3.4.

I have not included lists of exercises, but I will ask you various questions in the course of this book. Try to answer them. They are not difficult, but they aim at testing the level of your understanding.

Besides Sections 1.3–1.5, the core material is contained in Chapter 2 and the first sections of Chapter 3. Chapter 2 provides the construction of the Itô integral and a heuristic derivation of the Itô lemma, the chain rule of stochastic calculus. In Chapter 3 you will learn how to solve some simple stochastic differential equations. Section 3.3 on linear stochastic differential equation is mainly included in order to exercise the use of the Itô lemma. Section 3.4 will be interesting to those who want to visualize solutions to stochastic differential equations.

Chapter 4 is for those readers who want to see how stochastic calculus enters financial applications. Prior knowledge of economic theory is not required, but we will introduce a minimum of economic terminology which can be understood by everybody. If you can read through Section 4.1 on option pricing without major difficulties as regards stochastic calculus, you will have passed the examination on this course on elementary stochastic calculus.

At the end of this book you may want to know more about stochastic calculus and its applications. References to more advanced literature are given in the Notes and Comments at the end of each section. These references are not exhaustive; they do not include the theoretically most advanced textbook treatments, but they can be useful for the continuation of your studies.

> **You are now ready to start. Good luck!**

T.M.

1

Preliminaries

In this chapter we collect some basic facts needed for defining stochastic integrals. At a first reading, most parts of this chapter can be skipped, provided you have some basic knowledge of probability theory and stochastic processes. You may then want to start with Chapter 2 on Itô stochastic calculus and recall some facts from this chapter if necessary.

In Section 1.1 we recall elementary notions from probability theory such as *random variable, random vector, distribution, distribution function, density, expectation, moment, variance* and *covariance*. This small review cannot replace a whole course on probability, and so you are well recommended to consult your old lecture notes or a standard textbook. Section 1.2 is about *stochastic processes*. A stochastic process is a natural model for describing the evolution of real-life processes, objects and systems in time and space. One particular stochastic process plays a central rôle in this book: *Brownian motion*. We introduce it in Section 1.3 and discuss some of its elementary properties, in particular the non-differentiability and the unbounded variation of its sample paths. These properties indicate that Brownian sample paths are very irregular, and therefore a new, stochastic calculus has to be introduced for integrals with respect to Brownian motion.

In Section 1.4 we shortly review *conditional expectations*. Their precise definition is based on a deep mathematical theory, and therefore we only give some intuition on this concept. The same remark applies to Section 1.5, where we introduce an important class of stochastic processes: the *martingales*. It includes Brownian motion and indefinite Itô integrals as particular examples.

1.1 Basic Concepts from Probability Theory

1.1.1 Random Variables

The outcome of an experiment or game is random. A simple example is coin tossing: the possible outcomes "head" or "tail" are not predictable in the sense that they appear according to a random mechanism which is determined by the physical properties of the coin. A more complicated experiment is the stock market. There the random outcomes of the brokers' activities (which actually represent economic tendencies, political interests and their own instincts) are for example share prices and exchange rates. Another game is called "competition" and can be watched where products are on sale: the price of 1 kg bananas, say, is the outcome of a game between the shop owners, on the one hand, and between the shop owners and the customers, on the other hand.

The scientific treatment of an experiment requires that we assign a number to each random outcome. When tossing a coin, we can write "1" for "head" and "0" for "tail". Thus we get a *random variable* $X = X(\omega) \in \{0, 1\}$, where ω belongs to the *outcome space* $\Omega = \{\text{head,tail}\}$. The value of a share price of stock is already a random number, and so is the banana price in a greengrocers. These numbers $X(\omega)$ provide us with information about the experiment, even if we do not know who plays the game or what drives it.

Mathematicians make a clear cut between reality and a mathematical model: they define an abstract space Ω collecting all possible outcomes ω of the underlying experiment. It is an abstract space, i.e. it does not really matter what the ωs are. In mathematical language, the *random variable* $X = X(\omega)$ is nothing but a real-valued function defined on Ω.

The next step in the process of abstraction from reality is the probabilistic description of the random variable X:

Which are the most likely values $X(\omega)$, what are they concentrated around, what is their spread?

To approach these problems, one first collects "good" subsets of Ω, the *events*, in a class $\mathcal{F}$, say. In advanced textbooks $\mathcal{F}$ is called a *σ-field* or *σ-algebra*; see p. 62 for a precise definition. Such a class is supposed to contain all interesting events. What could $\mathcal{F}$ be for coin tossing? Certainly, $\{\omega : X(\omega) = 0\} = \{\text{tail}\}$ and $\{\omega : X(\omega) = 1\} = \{\text{head}\}$ must belong to $\mathcal{F}$, but also the union, difference, intersection of any events in $\mathcal{F}$, the set $\Omega = \{\text{head,tail}\}$ and its complement, the empty set $\emptyset$. This is a trivial example, but it shows what $\mathcal{F}$ should be like: if $A \in \mathcal{F}$, so is its complement A^c, and if $A, B \in \mathcal{F}$, so are $A \cap B$, $A \cup B$, $A \cup B^c$, $B \cup A^c$, $A \cap B^c$, $B \cap A^c$, etc.

If we consider a share price X, not only the events $\{\omega : X(\omega) = c\}$ should belong to $\mathcal{F}$, but also

$$\{\omega : a < X(\omega) \leq b\}, \quad \{\omega : b < X(\omega)\}, \quad \{\omega : X(\omega) \leq a\},$$

and many more events which can be relevant in this or that situation. As in the case of coin tossing, we would like that elementary operations such as $\cap, \cup,^c$ on the events of $\mathcal{F}$ should not lead outside the class $\mathcal{F}$. This is the intuitive meaning of a σ-field $\mathcal{F}$.

Probability, Distribution and Distribution Function

Now, where do the *probabilities* come in? When flipping a coin, "head" or "tail" occurs. Probabilities measure the likelihood that these events happen. If the coin is "fair" we assign the probability 0.5 to both events, i.e. $P(\{\omega : X(\omega) = 0\}) = P(\{\omega : X(\omega) = 1\}) = 0.5$. This mathematical definition is based on empirical evidence: if we flip a fair coin a large number of times, we expect that about 50% of the outcomes are heads and about 50% are tails. In probability theory, the *law of large numbers* gives the theoretical justification for this empirical observation.

This elementary example explains what a *probability measure* on the class $\mathcal{F}$ of the events is: to each event $A \in \mathcal{F}$ it assigns a number $P(A) \in [0,1]$. This number is the expected fraction of occurrences of the event A in a long series of experiments where A or A^c are observed.

Some elementary properties of probability measures are easily summarized:

For events $A, B \in \mathcal{F}$,

$$P(A \cup B) = P(A) + P(B) - P(A \cap B),$$

and, if A and B are disjoint,

$$P(A \cup B) = P(A) + P(B).$$

Moreover,

$$P(A^c) = 1 - P(A), \quad P(\Omega) = 1 \quad \text{and} \quad P(\emptyset) = 0.$$

The relationship between random variables and probability can be characterized by certain numerical quantities. In what follows, we consider some of them.

The collection of the probabilities

$$F_X(x) = P(X \leq x) = P(\{\omega : X(\omega) \leq x\}), \quad x \in \mathbb{R} = (-\infty, \infty),$$

is the *distribution function F_X* of X.

It yields the probability that X belongs to the interval $(a, b]$. Indeed,

$$P(\{\omega : a < X(\omega) \leq b\}) = F_X(b) - F_X(a), \quad a < b.$$

Moreover, we also obtain the probability that X is equal to a number:

$$P(X = x)$$

$$= \quad P(\{\omega : X(\omega) = x\}) = P(\{\omega : X \leq x\}) - P(\{\omega : X < x\})$$

$$= \quad P(\{\omega : X(\omega) \leq x\}) - \lim_{h \downarrow 0} P(\{\omega : X(\omega) \leq x - h\})$$

$$= \quad F_X(x) - \lim_{h \downarrow 0} F_X(x - h).$$

With these probabilities one can approximate the probability of the event $\{\omega : X(\omega) \in B\}$ for very complicated subsets B of $\mathbb{R}$.

The collection of the probabilities

$$P_X(B) = P(X \in B) = P(\{\omega : X(\omega) \in B\})$$

for suitable subsets $B \subset \mathbb{R}$ is the *distribution* of X.

"Suitable" subsets of $\mathbb{R}$ are the so-called *Borel sets*. They are obtained by a countable number of operations $\cap$, $\cup$ or ^{c} acting on the intervals; see p. 64 for a precise definition.

The distribution P_X and the distribution function F_X are equivalent notions in the sense that both of them can be used to calculate the probability of any event $\{X \in B\}$.

A distribution function is continuous or it has jumps. We first consider the special case when the distribution function F_X is a pure jump function:

$$F_X(x) = \sum_{k:\, x_k \le x} p_k, \quad x \in \mathbb{R}, \tag{1.1}$$

where

$$0 \le p_k \le 1 \quad \text{for all } k \text{ and } \sum_{k=1}^{\infty} p_k = 1.$$

The distribution function (1.1) and the corresponding distribution are said to be *discrete*; a random variable with distribution function (1.1) is a *discrete random variable*.

A discrete random variable only assumes a finite or countably infinite number of values $x_1, x_2, \ldots$, and $p_k = P(X = x_k)$. In particular, the distribution function F_X has an upward jump of size p_k at $x = x_k$. For example, the random variable X related to coin tossing is discrete: it assumes the values 0 and 1. The price for any product on sale in a supermarket is a discrete random variable: it may assume the values $0, 0.01, 0.02, \ldots$ \$, say.

Example 1.1.1 (Two important discrete distributions)
Important discrete distributions are the *binomial distribution* $Bin(n, p)$ with parameters $n \in \mathbb{N} = \{0, 1, 2, \ldots\}$ and $p \in (0, 1)$:

$$P(X = k) = \binom{n}{k} p^k (1-p)^{n-k}, \quad k = 0, 1, \ldots, n,$$

and the *Poisson distribution* $Poi(\lambda)$ with parameter $\lambda > 0$:

$$P(X = k) = e^{-\lambda} \frac{\lambda^k}{k!}, \quad k = 0, 1, 2, \ldots.$$

See Figure 1.1.2 for an illustration. $\qquad \square$

In contrast to discrete distributions and random variables, the distribution function of a *continuous random variable* does not have jumps, hence $P(X = x) = 0$ for all x, or equivalently,

$$\lim_{h \to 0} F_X(x + h) = F_X(x) \quad \text{for all } x, \tag{1.2}$$

i.e. such a random variable assumes any particular value with probability 0. A continuous random variable gains its name from the continuity property (1.2) of the distribution function F_X.

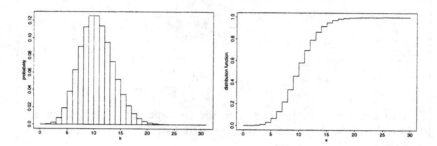

Figure 1.1.2 Left: *the probabilities $P(X = k)$, $k = 0, 1, 2, \ldots$, of the Poisson distribution with parameter $\lambda = 10$.* Right: *the corresponding distribution function.*

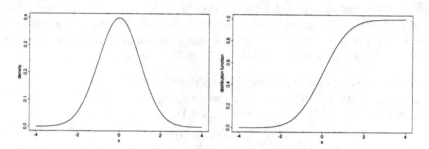

Figure 1.1.3 Left: *the density of the standard normal distribution (mean 0, variance 1).* Right: *the corresponding distribution function.*

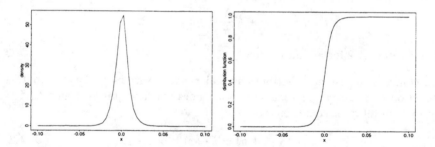

Figure 1.1.4 Left: *the density of the log-returns $X_t = \ln Y_t - \ln Y_{t-1}$ of the daily closing prices Y_t of the S&P index. The S&P is one of the basic US industrial indices.* Right: *the corresponding distribution function. A comparison with Figure 1.1.3 indicates that the latter distribution is certainly not normal.*

> Most continuous distributions of interest have a *density* f_X:
>
> $$F_X(x) = \int_{-\infty}^{x} f_X(y)\, dy, \quad x \in \mathbb{R},$$
>
> where
>
> $$f_X(x) \geq 0 \quad \text{for every } x \in \mathbb{R} \text{ and } \int_{-\infty}^{\infty} f_X(y)\, dy = 1.$$

Example 1.1.5 (The normal and uniform distributions)
An important continuous distribution is the *normal* or *Gaussian distribution* $N(\mu, \sigma^2)$ with parameters $\mu \in \mathbb{R}$, $\sigma^2 > 0$. It has density

$$f_X(x) = \frac{1}{\sqrt{2\pi}\sigma} \exp\left\{ -\frac{(x-\mu)^2}{2\sigma^2} \right\}, \quad x \in \mathbb{R}. \tag{1.3}$$

If X is $N(0,1)$ (standard normal) we write φ for the density f_X and Φ for the distribution function F_X. For an illustration of the standard normal density and the corresponding distribution function, see Figure 1.1.3.
The *uniform* distribution $U(a,b)$ on (a,b) has density

$$f_X(x) = \begin{cases} \dfrac{1}{b-a}, & \text{if } x \in (a,b), \\ 0, & \text{otherwise}. \end{cases} \qquad \square$$

The value of an exchange rate or share price can, at least theoretically, assume any positive real number. Clearly, there are technical limitations: a computer or pocket calculator is not capable of saving the value of an exchange rate with infinitely many digits, $\sqrt{2}$ say; every number in the computer's memory is rounded off. Therefore any random variable of practical interest is actually discrete.... However, it is often convenient to think of such a variable as a continuous one. There may be theoretical reasons. For example, the normal distribution appears as a limit distribution via the central limit theorem; see p. 45. Many functions of a sample are therefore approximately normal, hence their limit distribution is continuous. But there are also practical reasons: it is often less tedious to work with a well-studied continuous distribution (such as the normal, exponential, gamma, uniform) because we can use standard knowledge (in particular, standard software packages) about its density, moments, quantiles, etc., and we can possibly obtain some nice explicit expressions for these quantities.

Expectation, Variance and Moments

Interesting characteristics of a random variable X are the *expectation EX*, the *variance* $\text{var}(X)$ and the *moments* $E(X^l)$.

The *expectation* or *mean value* of a random variable X with density f_X is given by

$$\mu_X = EX = \int_{-\infty}^{\infty} x f_X(x) \, dx \, .$$

The *variance* of X is defined as

$$\sigma_X^2 = \text{var}(X) = \int_{-\infty}^{\infty} (x - \mu_X)^2 f_X(x) \, dx \, .$$

The *l*th *moment* of X for $l \in \mathbb{N}$ is defined as

$$E(X^l) = \int_{-\infty}^{\infty} x^l f_X(x) \, dx \, .$$

For a real-valued function g the expectation of $g(X)$ is given by

$$Eg(X) = \int_{-\infty}^{\infty} g(x) f_X(x) \, dx \, .$$

The *expectation* or *mean value* of a discrete random variable X with probabilities $p_k = P(X = x_k)$ is given by

$$\mu_X = EX = \sum_{k=1}^{\infty} x_k \, p_k \, .$$

The *variance* of X is defined as

$$\sigma_X^2 = \text{var}(X) = \sum_{k=1}^{\infty} (x_k - \mu_X)^2 \, p_k \, .$$

The *l*th *moment* of X for $l \in \mathbb{N}$ is defined as

$$E(X^l) = \sum_{k=1}^{\infty} x_k^l \, p_k \, .$$

For a real-valued function g the expectation of $g(X)$ is given by

$$Eg(X) = \sum_{k=1}^{\infty} g(x_k)\, p_k\,.$$

We can regard the expectation μ_X as the "center of gravity" of the random variable X, i.e. the *random* values $X(\omega)$ are concentrated around the *non-random* number μ_X. The expectation is quite often taken as a surrogate for the size of the random variable. For example, it is a simple means for predicting the future values in a time series.

The spread or dispersion of the random values $X(\omega)$ around the expectation μ_X is described by the variance

$$
\begin{aligned}
\sigma_X^2 &= \operatorname{var}(X) = E(X - \mu_X)^2 \\
&= E(X^2 - 2\mu_X X + \mu_X^2) = E(X^2) - 2\mu_X^2 + \mu_X^2 \\
&= E(X^2) - \mu_X^2
\end{aligned}
$$

and the *standard deviation* σ_X.

Recall the normal density from (1.3). The parameter μ is the expectation μ_X and the parameter σ^2 is the variance σ_X^2 of a random variable X with density (1.3). It is a well-known fact (and easy to verify, for example with the computer) that for an $N(\mu, \sigma^2)$ random variable X,

$$P(\mu - 1.96\,\sigma \le X \le \mu + 1.96\,\sigma)$$

$$= \Phi(\mu + 1.96\,\sigma) - \Phi(\mu - 1.96\,\sigma) = 0.95\,. \tag{1.4}$$

So there is a 95% chance that the normal random variable X assumes values in $[\mu - 1.96\,\sigma, \mu + 1.96\,\sigma]$. Analogously to (1.4) one can formulate a heuristic 2σ-rule (it is nothing but this: one can construct counterexamples) that for a "nice" random variable X the probability

$$P(\mu_X - 2\,\sigma_X \le X \le \mu_X + 2\,\sigma_X)$$

is close to 1. This rule is also supported by the *Chebyshev inequality*

$$P(|X - \mu_X| > x) \le x^{-2}\sigma_X^2\,, \quad x > 0\,,$$

which provides us with a sufficiently accurate bound of the probability that the absolute deviation of the random variable X from its expectation exceeds some threshold x.

1.1.2   Random Vectors

In what follows we frequently make use of finite-dimensional and infinite-dimensional random structures. We commence with finite-dimensional random vectors as a first step toward the definition of a stochastic process.

$\mathbf{X} = (X_1, \ldots, X_n)$ is an n-dimensional *random vector* if its components $X_1, \ldots, X_n$ are one-dimensional real-valued random variables.

If we interpret $t = 1, \ldots, n$ as equidistant instants of time, X_t can stand for the outcome of an experiment at time t. Such a *time series* may, for example, consist of BMW share prices X_t at n succeeding days. Clearly, t is "mathematical time", thus nothing but an index or a counting variable. For example, a random vector can describe the state of the weather in Wellington (NZ) at a given time: X_1 could be the temperature, X_2 the air pressure and X_3 the windspeed (the latter is usually close to infinity!).

Analogously to one-dimensional random variables one can introduce the distribution function, the expectation, moments and the covariance matrix of a random vector in order to describe its distribution and its dependence structure. The latter aspect is a new one; dependence does not make sense when we talk just about one random variable.

Probability, Distribution and Distribution Function

Toss a fair coin twice. We consider the four pairs (H,H), (T,T), (H,T) and (T,H) (H=head, T=tail) as outcomes of the experiment "flipping a coin twice". These four pairs constitute the outcome space Ω. As before, we assign 1 to H and 0 to T. In this way we obtain two random variables X_1 and X_2, and $\mathbf{X} = (X_1, X_2)$ is a two-dimensional random vector. Notice that

$$\mathbf{X}(H,H) = (1,1)\,,\ \mathbf{X}(T,T) = (0,0)\,,\ \mathbf{X}(T,H) = (0,1)\,,\ \mathbf{X}(H,T) = (1,0)\,.$$

If the coin is indeed fair, we can assign the probability 0.25 to each of the four outcomes, i.e.

$$P(\{\omega : \mathbf{X}(\omega) = (k,l)\}) = 0.25\,, \quad k,l \in \{0,1\}\,.$$

As before, we consider a collection $\mathcal{F}$ of subsets of Ω and define a probability measure on it, i.e. we assign a number $P(A) \in [0,1]$ to each $A \in \mathcal{F}$.

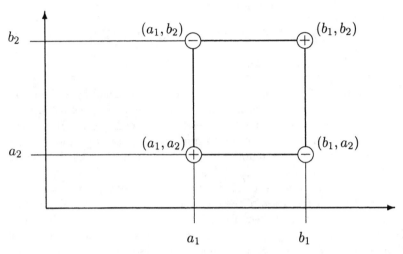

Figure 1.1.6

The collection of the probabilities

$$F_{\mathbf{X}}(\mathbf{x}) = P(X_1 \leq x_1, \dots, X_n \leq x_n) \qquad (1.5)$$
$$= P(\{\omega : X_1(\omega) \leq x_1, \dots, X_n(\omega) \leq x_n\}),$$
$$\mathbf{x} = (x_1, \dots, x_n) \in \mathbb{R}^n,$$

is the *distribution function* $F_{\mathbf{X}}$ of $\mathbf{X}$.

It provides us with the probability of the event that $\mathbf{X}$ assumes values in the rectangle

$$(\mathbf{a}, \mathbf{b}] = \{\mathbf{x} : a_i < x_i \leq b_i, \quad i = 1, \dots, n\}.$$

For example, if $\mathbf{X}$ is two-dimensional,

$$P(\mathbf{X} \in (\mathbf{a}, \mathbf{b}]) = F_{\mathbf{X}}(b_1, b_2) + F_{\mathbf{X}}(a_1, a_2) - F_{\mathbf{X}}(a_1, b_2) - F_{\mathbf{X}}(b_1, a_2).$$

Check that this formula is correct; see also Figure 1.1.6. As in the case of one-dimensional random variables, these probabilities approximate $P(\mathbf{X} \in B)$ for very general sets B.

The collection of the probabilities

$$P_{\mathbf{X}}(B) = P(\mathbf{X} \in B) = P(\{\omega : \mathbf{X}(\omega) \in B\})$$

for suitable subsets $B \subset \mathbb{R}^n$ constitutes the *distribution* of $\mathbf{X}$.

"Suitable" subsets of $\mathbb{R}^n$ are the *Borel sets* which are obtained by a countable number of operations $\cup$, $\cap$ or ^{c} acting on the intervals in $\mathbb{R}^n$; see p. 64 for a precise definition. For example, point sets, balls and rectangles are Borel sets. In a mathematical sense, the distribution and the distribution function of a random vector $\mathbf{X}$ are equivalent notions. Both, $F_{\mathbf{X}}$ and $P_{\mathbf{X}}$, can be used to calculate the probability of any event $\{\mathbf{X} \in B\}$.

Notice that the distribution of $\mathbf{X} = (X_1, \ldots, X_n)$ contains the whole information about the distribution of the components X_i, pairs (X_i, X_j), triples (X_i, X_j, X_k), etc. This is easily seen from (1.5): you get the distribution function of X_1 by formally setting $x_2 = \cdots = x_n = \infty$, the distribution function of (X_1, X_2) by setting $x_3 = \cdots = x_n = \infty$, etc.

Analogously to random variables one can introduce discrete and continuous random vectors and distributions. For our purposes, continuous random vectors with a density will be relevant, and so we mostly restrict our attention to them.

If the distribution of a random vector $\mathbf{X}$ has *density* $f_{\mathbf{X}}$, one can represent the distribution function $F_{\mathbf{X}}$ of $\mathbf{X}$ as

$$F_{\mathbf{X}}(x_1, \ldots, x_n) = \int_{-\infty}^{x_1} \cdots \int_{-\infty}^{x_n} f_{\mathbf{X}}(y_1, \ldots, y_n) \, dy_1 \cdots dy_n \,,$$

$$(x_1, \ldots, x_n) \in \mathbb{R}^n \,,$$

where the density is a function satisfying

$$f_{\mathbf{X}}(\mathbf{x}) \geq 0 \quad \text{for every } \mathbf{x} \in \mathbb{R}^n$$

and

$$\int_{-\infty}^{\infty} \cdots \int_{-\infty}^{\infty} f_{\mathbf{X}}(y_1, \ldots, y_n) \, dy_1 \cdots dy_n = 1 \,.$$

If a vector $\mathbf{X}$ has density $f_{\mathbf{X}}$, all its components X_i, the vectors of the pairs (X_i, X_j), triples (X_i, X_j, X_k), etc., have a density. They are called *marginal densities*.

Example 1.1.7 (Marginal densities: the case $n = 3$)
We consider the case $n = 3$. Then the marginal densities are obtained as follows:

$$f_{X_1}(x_1) \;=\; \int_{-\infty}^{\infty} \int_{-\infty}^{\infty} f_{\mathbf{X}}(\mathbf{x})\, dx_2 dx_3\,, \quad f_{X_1,X_2}(x_1,x_2) = \int_{-\infty}^{\infty} f_{\mathbf{X}}(\mathbf{x})\, dx_3\,,$$

$f_{X_2}(x_2)$ is obtained by integrating $f_{\mathbf{X}}(\mathbf{x})$ with respect to x_1 and x_3, f_{X_1,X_3} by integrating $f_{\mathbf{X}}(\mathbf{x})$ with respect to x_2, etc.   □

One case is particularly simple: if the density $f_{\mathbf{X}}(\mathbf{x})$ can be written as a product of non-negative functions g_i:

$$f_{\mathbf{X}}(\mathbf{x}) = g_1(x_1)\cdots g_n(x_n)\,, \quad \mathbf{x} \in \mathbb{R}^n\,.$$

In this case, $\int_{-\infty}^{\infty} g_i(x_i)\, dx_i = 1$ for $i = 1,\ldots,n$, i.e. the functions $g_i(x_i)$ are one-dimensional probability densities, and then necessarily

$$f_{X_i}(x_i) = g_i(x_i)\,, \quad f_{X_i,X_j}(x_i,x_j) = g_i(x_i)\, g_j(x_j)\,, \quad \text{etc.}$$

Verify this!

Example 1.1.8 (Gaussian random vector)
A *Gaussian* or *normal random vector* has a Gaussian or normal distribution. The *n-dimensional normal* or *Gaussian distribution* is given by its density

$$f_{\mathbf{X}}(\mathbf{x}) = \frac{1}{(2\pi)^{n/2}(\det \Sigma)^{1/2}} \exp\left\{-\frac{1}{2}(\mathbf{x}-\mu)\Sigma^{-1}(\mathbf{x}-\mu)'\right\}\,, \quad \mathbf{x} \in \mathbb{R}^n\,, \quad (1.6)$$

with parameters $\mu \in \mathbb{R}^n$ and Σ. (Here and in what follows, $\mathbf{y}'$ denotes the transpose of the vector $\mathbf{y}$; see p. 211.) The quantity Σ is a symmetric positive-definite $n \times n$ matrix, Σ^{-1} is its inverse and $\det \Sigma$ its determinant. See Figure 1.1.10 for an illustration in the case $n = 2$.   □

Expectation, Variance and Covariance

The expectation of a random vector has a similar function as the mean value of a random variable. The values $\mathbf{X}(\omega)$ are concentrated around it.

The *expectation* or *mean value* of a random vector $\mathbf{X}$ is given by

$$\mu_{\mathbf{X}} = E\mathbf{X} = (EX_1,\ldots,EX_n)\,.$$

The *covariance matrix* of **X** is defined as

$$\Sigma_{\mathbf{X}} = (\text{cov}(X_i, X_j)\,;\, i, j = 1, \ldots, n)\,,$$

where

$$
\begin{aligned}
\text{cov}(X_i, X_j) &= E[(X_i - \mu_{X_i})(X_j - \mu_{X_j})] \\
&= E(X_i X_j) - \mu_{X_i}\mu_{X_j}\,,
\end{aligned}
$$

is the *covariance of X_i and X_j*. Notice that $\text{cov}(X_i, X_i) = \sigma_{X_i}^2$.

Example 1.1.9 (Continuation of Example 1.1.8)
Recall from (1.6) the density of a multivariate Gaussian random vector **X**. The parameter μ is the expectation $\mu_{\mathbf{X}}$ of **X**, and Σ is its covariance matrix $\Sigma_{\mathbf{X}}$. Thus the density of a Gaussian vector (hence its distribution) is completely determined via its expectation and covariance matrix. In particular, if $\mu = \mathbf{0}$ and Σ is the n-dimensional identity matrix I_n, we have $\det I_n = 1$ and $\Sigma^{-1} = I_n$. The density $f_{\mathbf{X}}$ is then simply the product of n standard normal densities:

$$f_{\mathbf{X}}(x_1, \ldots, x_n) = \varphi(x_1)\,\cdots\,\varphi(x_n)\,.$$

We write $N(\mu, \Sigma)$ for the distribution of an n-dimensional Gaussian vector **X** with expectation μ and covariance matrix Σ. Such a vector has the appealing property that it remains Gaussian under linear transformations (recall that μ' and A' denote the transposes of μ and A, respectively):

Let $\mathbf{X} = (X_1, \ldots, X_n)$ have an $N(\mu, \Sigma)$ distribution and A be an $m \times n$ matrix. Then $A\mathbf{X}'$ has an $N(A\mu', A\Sigma A')$ distribution.

$\square$

It is convenient to standardize covariances by dividing the corresponding random variables by their standard deviations. The resulting quantity

$$
\begin{aligned}
\text{corr}(X_1, X_2) &= \frac{\text{cov}(X_1, X_2)}{\sigma_{X_1}\sigma_{X_2}} \\
&= \frac{E[(X_1 - \mu_{X_1})(X_2 - \mu_{X_2})]}{\sigma_{X_1}\sigma_{X_2}}
\end{aligned}
$$

is the *correlation of X_1 and X_2*. As a result of this standardization, the correlation of two random variables is always between -1 and $+1$. Check this fact by an application of the Cauchy–Schwarz inequality; see p. 188.

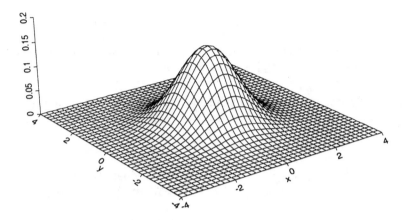

Figure 1.1.10 *Density of the 2-dimensional standard normal density ($\mu = 0$ and $\Sigma = I_2$ is the identity matrix).*

1.1.3 Independence and Dependence

Flip a fair coin twice and let the random numbers $X_1(\omega), X_2(\omega) \in \{0,1\}$ be the corresponding outcomes of the first and second experiment. It is easy to verify that

$$P(X_1 = k,\, X_2 = l) = P(X_1 = k)\, P(X_2 = l), \quad k, l \in \{0,1\}.$$

This property is called *independence* of the random variables X_1 and X_2. Intuitively, independence means that the first experiment does not influence the second one, and vice versa. For example, knowledge of X_1 does not allow one to predict the value of X_2, and vice versa.

Below we recall some of the essential definitions and properties of independent events and independent random variables.

Two events A_1 and A_2 are *independent* if

$$P(A_1 \cap A_2) = P(A_1)\, P(A_2).$$

Two random variables X_1 and X_2 are *independent* if

$$P(X_1 \in B_1 \, , \, X_2 \in B_2) = P(X_1 \in B_1) \, P(X_2 \in B_2)$$

for all suitable subsets B_1 and B_2 of $\mathbb{R}$. This means that the events $\{X_1 \in B_1\}$ and $\{X_2 \in B_2\}$ are independent.

Alternatively, one can define independence via distribution functions and densities. The random variables X_1 and X_2 are independent if and only if

$$F_{X_1, X_2}(x_1, x_2) = F_{X_1}(x_1) \, F_{X_2}(x_2) \, , \quad x_1 \, , \, x_2 \in \mathbb{R} \, .$$

Assume that (X_1, X_2) has density f_{X_1, X_2} with marginal densities f_{X_1} and f_{X_2}; see p. 16. Then the random variables X_1 and X_2 are independent if and only if

$$f_{X_1, X_2}(x_1, x_2) = f_{X_1}(x_1) \, f_{X_2}(x_2) \, , \quad x_1 \, , \, x_2 \in \mathbb{R} \, .$$

The definition of independence can be extended to an arbitrary finite number of events and random vectors. Notice that independence of the components of a random vector implies the independence of each pair of its components, but the converse is in general not true.

The events $A_1, \ldots, A_n$ are *independent* if, for every choice of indices $1 \leq i_1 < \cdots < i_k \leq n$ and integers $1 \leq k \leq n$,

$$P(A_{i_1} \cap \cdots \cap A_{i_k}) = P(A_{i_1}) \cdots P(A_{i_k}) \, .$$

The random variables $X_1, \ldots, X_n$ are *independent* if, for every choice of indices $1 \leq i_1 < \cdots < i_k \leq n$, integers $1 \leq k \leq n$ and all suitable subsets $B_1, \ldots, B_n$ of $\mathbb{R}$,

$$P(X_{i_1} \in B_{i_1} \, , \ldots , X_{i_k} \in B_{i_k}) = P(X_{i_1} \in B_{i_1}) \cdots P(X_{i_k} \in B_{i_k}) \, .$$

This means that the events $\{X_1 \in B_1\}, \ldots, \{X_n \in B_n\}$ are independent.

The random variables $X_1, \ldots, X_n$ are independent if and only if their joint distribution function can be written as follows:

$$F_{X_1, \ldots, X_n}(x_1, \ldots, x_n) = F_{X_1}(x_1) \cdots F_{X_n}(x_n) \, , \quad (x_1, \ldots, x_n) \in \mathbb{R}^n \, .$$

If the random vector $\mathbf{X} = (X_1, \ldots, X_n)$ has density $f_{\mathbf{X}}$, then $X_1, \ldots, X_n$ are independent if and only if

$$f_{X_1,\ldots,X_n}(x_1,\ldots,x_n) = f_{X_1}(x_1) \cdots f_{X_n}(x_n), \quad (x_1,\ldots,x_n) \in \mathbb{R}^n. \quad (1.7)$$

Example 1.1.11 (Continuation of Example 1.1.9)
Recall from (1.6) the density of an n-dimensional Gaussian vector $\mathbf{X}$. It is easily seen from the form of the density that its components are independent if and only if the covariance matrix Σ is diagonal. This means that $\mathrm{corr}(X_i, X_j) = \mathrm{cov}(X_i, X_j) = 0$ for $i \neq j$, and so we can write the density of $\mathbf{X}$ in the form (1.7). Thus, *in the Gaussian case, uncorrelatedness and independence are equivalent notions.* This statement is wrong for non-Gaussian random vectors; see Example 1.1.12. □

An important consequence of the independence of random variables is the following property:

If $X_1, \ldots, X_n$ are independent, then for any real-valued functions $g_1, \ldots, g_n$,

$$E[g_1(X_1) \cdots g_n(X_n)] = Eg_1(X_1) \cdots Eg_n(X_n),$$

provided the considered expectations are well defined.

In particular, we may conclude that the independent random variables X_1 and X_2 are *uncorrelated*, i.e. $\mathrm{corr}(X_1, X_2) = \mathrm{cov}(X_1, X_2) = 0$. The converse is in general not true.

Example 1.1.12 (Uncorrelated random variables are not necessarily independent.)
Let X be a standard normal random variable. Since X is symmetric (i.e. X and $-X$ have the same distribution), so is X^3, and therefore both X and X^3 have expectation zero. Thus

$$\mathrm{cov}(X, X^2) = E(X^3) - EX\,E(X^2) = 0,$$

but X and X^2 are clearly dependent: since $\{X \in [-1,1]\} = \{X^2 \in [0,1]\}$, we obtain

$$
\begin{aligned}
P(X \in [-1,1],\, X^2 \in [0,1]) \quad &= \quad P(X \in [-1,1]) \\
> \quad P(X \in [-1,1])P(X^2 \in [0,1]) \quad &= \quad [P(X \in [-1,1])]^2.
\end{aligned}
$$
□

Example 1.1.13 (Autocorrelations of a time series)
For a time series $X_0, X_1, X_2, \ldots$ the autocorrelation at lag h is defined by $\mathrm{corr}(X_0, X_h)$, $h = 0, 1, \ldots$. A claim which can frequently be found in the literature is that financial time series (derived from stock indices, share prices, exchange rates, etc.) are nearly uncorrelated. This is supported by the sample autocorrelations of the daily log-returns X_t of the S&P index; see Figure 1.1.14. In contrast to this observation, the estimated autocorrelations of the absolute values $|X_t|$ are different from zero even for large lags h. This indicates that there is dependence in this time series.   □

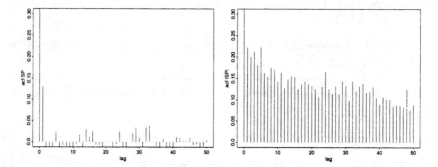

Figure 1.1.14 *The estimated autocorrelations of the S&P index* (left) *and of its absolute values* (right)*; see Example 1.1.13; cf. the comments in Figure 1.1.4.*

In what follows, we will often deal with infinite collections $(X_t, t \in T)$ of random variables X_t, i.e. T is an infinite index set. In this set-up, we may also introduce independence:

The collection of random variables $(X_t, t \in T)$ is *independent* if for every choice of distinct indices $t_1, \ldots, t_n \in T$ and $n \geq 1$ the random variables $X_{t_1}, \ldots, X_{t_n}$ are independent. This collection is *independent and identically distributed* (*iid*) if it is independent and all random variables X_t have the same distribution.

Notes and Comments

In this section we recalled some elementary probability theory which can be found in every textbook on the topic; see for instance Pitman (1993) for an elementary level and Gut (1995) for an intermediate course. Also many text-books on statistics often begin with an introduction to probability theory; see for example Mendenhall, Wackerly and Scheaffer (1990).

1.2   Stochastic Processes

We suppose that the exchange rate NZ\$/US\$ at every fixed instant t between 9 a.m. and 10 a.m. this morning is random. Therefore we can interpret it as a realization $X_t(\omega)$ of the random variable X_t, and so we observe $X_t(\omega)$, $9 \le t \le 10$. In order to make a guess at 10 a.m. about the exchange rate $X_{11}(\omega)$ at 11 a.m. it is reasonable to look at the whole evolution of $X_t(\omega)$ between 9 a.m. and 10 a.m. This is also a demand of the high standard technical devices which provide us with almost continuous information about the process considered. A mathematical model for describing such a phenomenon is called a stochastic process.

A *stochastic process* X is a collection of random variables

$$(X_t, t \in T) = (X_t(\omega), t \in T, \omega \in \Omega),$$

defined on some space Ω.

For our purposes, T is often an interval, for example $T = [a, b]$, $[a, b)$ or $[a, \infty)$ for $a < b$. Then we call X a *continuous-time* process in contrast to *discrete-time* processes. In the latter case, T is a finite or countably infinite set. For obvious reasons, the index t of the random variable X_t is frequently referred to as *time*, and we will follow this convention.

A stochastic process X is a function of two variables.

For a fixed instant of time t, it is a random variable:

$$X_t = X_t(\omega), \quad \omega \in \Omega.$$

For a fixed random outcome $\omega \in \Omega$, it is a function of time:

$$X_t = X_t(\omega), \quad t \in T.$$

This function is called a *realization*, a *trajectory* or a *sample path* of the process X.

These two aspects of a stochastic process are illustrated in Figure 1.2.1.

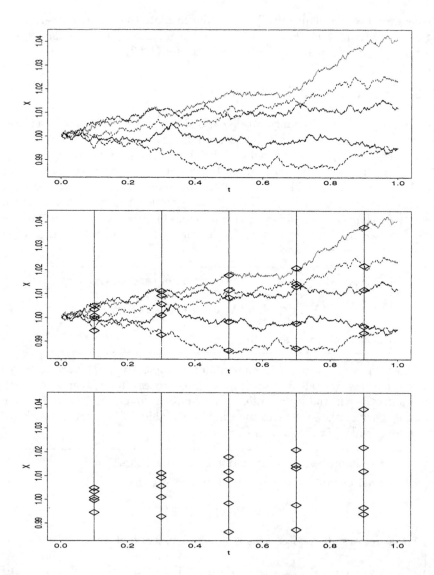

Figure 1.2.1 5 *sample paths of a stochastic process* $(X_t, t \in [0,1])$. Top: *every path corresponds to a different* $\omega \in \Omega$. Middle and bottom: *the values on the vertical line at* $t = 0.1, \ldots, 0.9$ *visualize the random variables* $X_{0.1}, \ldots, X_{0.9}$; *they occur as the projections of the sample paths on the vertical lines.*

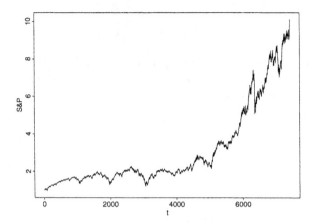

Figure 1.2.2 *The (scaled) daily values of the S&P index over a period of 7,422 days. The graph suggests that we consider the S&P time series as the sample path of a continuous-time process. If there are many values in a time series such that the instants of time $t \in T$ are "dense" in an interval, then one may want to interpret this discrete-time process as a continuous-time process. The sample paths of a real-life continuous-time process are always reported at discrete instants of time. Depending on the situation, one has to make a decision which model (discrete- or continuous-time) is more appropriate.*

Example 1.2.3 A *time series*

$$X_t, \quad t = 0, \pm 1, \pm 2, \ldots,$$

is a discrete-time process with $T = \mathbb{Z} = \{0, \pm 1, \pm 2, \ldots\}$. Time series constitute an important class of stochastic processes. They are relevant models in many applications, where one is interested in the evolution of a real-life process. Such series represent, for example, the daily body temperature of a patient in a hospital, the daily returns of a price or the monthly number of air traffic passengers in the US. The most popular theoretical time series models are the *ARMA* (**A**uto**R**egressive **M**oving **A**verage) processes. They are given by certain difference equations in which an iid sequence (Z_t) (see p. 22), the so-called *noise*, is involved. For example, a moving average of order $q \geq 1$ is defined as

$$X_t = Z_t + \theta_1 Z_{t-1} + \cdots + \theta_q Z_{t-q}, \quad t \in \mathbb{Z},$$

and an autoregressive process of order 1 is given by

$$X_t = \phi X_{t-1} + Z_t, \quad t \in \mathbb{Z}.$$

Here $\theta_1, \ldots, \theta_q$ and ϕ are given real parameters. Time series models can be understood as discretizations of stochastic differential equations. We will see this for the autoregressive process on p. 141.

Figure 1.2.4 shows two examples.   □

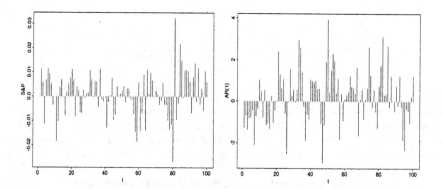

Figure 1.2.4 *Two time series X_t, $t = 1, \ldots, 100$. Left: 100 successive daily log-returns of the S&P index; see Figure 1.1.4. Right: a simulated sample path of the autoregressive process $X_t = 0.5X_{t-1} + Z_t$, where Z_t are iid $N(0,1)$ random variables; see Example 1.2.3.*

We see that the concepts of a random variable X and of a stochastic process $(X_t, t \in T)$ are not so much different. Both have random realizations, but the realization $X(\omega)$ of a random variable is a number, whereas the realization $X_t(\omega), t \in T$, of a stochastic process is a function on T. So we are completely correct if we understand a stochastic process to be a "random element" taking functions as values. Moreover, we can interpret a random variable and a random vector as special stochastic processes with a finite index set T.

Distribution

In analogy to random variables and random vectors we want to introduce non-random characteristics of a stochastic process such as its distribution, expectation, etc. and describe its dependence structure. This is a task much more complicated than the description of a random vector. Indeed, a non-trivial stochastic process $X = (X_t, t \in T)$ with infinite index set T is an infinite-dimensional object; it can be understood as the infinite collection of the random variables X_t, $t \in T$. Since the values of X are functions on T, the *distribution of X* should be defined on subsets of a certain "function space", i.e.

$$P(X \in A), \quad A \in \mathcal{F}, \tag{1.8}$$

where $\mathcal{F}$ is a collection of suitable subsets of this space of functions. This approach is possible, but requires advanced mathematics, and so we try to find some simpler means.

The key observation is that a stochastic process can be interpreted as a collection of random vectors.

The *finite-dimensional distributions* (*fidis*) of the stochastic process X are the distributions of the finite-dimensional vectors

$$(X_{t_1}, \ldots, X_{t_n}), \quad t_1, \ldots, t_n \in T,$$

for all possible choices of times $t_1, \ldots, t_n \in T$ and every $n \geq 1$.

We can imagine the fidis much easier than the complicated distribution (1.8) of a stochastic process. It can be shown that the fidis determine the distribution of X. In this sense, we refer to the collection of the fidis as the *distribution of the stochastic process*.

Stochastic processes can be classified according to different criteria. One of them is the kind of fidis.

Example 1.2.5 (Gaussian process)
Recall from (1.6) the definition of an n-dimensional Gaussian density. A stochastic process is called *Gaussian* if all its fidis are multivariate Gaussian. We learnt in Example 1.1.9 that the parameters μ and Σ of a Gaussian vector are its expectation and covariance matrix, respectively. Hence the distribution of a Gaussian stochastic process is determined only by the collection of the expectations and covariance matrices of the fidis.

A simple Gaussian process on $T = [0, 1]$ consists of iid $N(0, 1)$ random variables. In this case the fidis are characterized by the distribution functions

$$P(X_{t_1} \leq x_1, \ldots, X_{t_n} \leq x_n)$$
$$= P(X_{t_1} \leq x_1) \cdots P(X_{t_n} \leq x_n)$$
$$= \Phi(x_1) \cdots \Phi(x_n)$$

$$0 \leq t_1 \leq \cdots \leq t_n \leq 1, \quad (x_1, \ldots, x_n) \in \mathbb{R}^n.$$

The sample paths of this process are very irregular. See Figure 1.2.6 for an illustration.   □

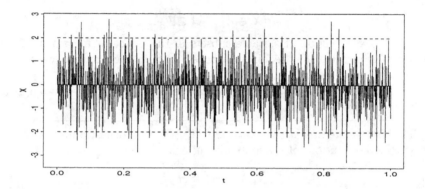

Figure 1.2.6 *A sample path of the Gaussian process* $(X_t, t \in [0,1])$*, where the* X_t*s are iid* $N(0,1)$*; see Example 1.2.5. The expectation function is* $\mu_X(t) = 0$ *and the dashed lines indicate the curves* $\pm 2\sigma_X(t) = \pm 2$*; see Example 1.2.7.*

Expectation and Covariance Function

For a random vector $\mathbf{X} = (X_1, \ldots, X_n)$ we defined the expectation $\mu_{\mathbf{X}} = (EX_1, \ldots, EX_n)$ and the covariance matrix $\Sigma_{\mathbf{X}} = (\text{cov}(X_i, X_j), i, j = 1, \ldots, n)$. A stochastic process $X = (X_t, t \in T)$ can be considered as the collection of the random vectors $(X_{t_1}, \ldots, X_{t_n})$ for $t_1, \ldots, t_n \in T$ and $n \geq 1$. For each of them we can determine the expectation and covariance matrix. Alternatively, we can consider these quantities as functions of $t \in T$:

The *expectation function* of X is given by

$$\mu_X(t) = \mu_{X_t} = EX_t, \quad t \in T.$$

The *covariance function* of X is given by

$$c_X(t, s) = \text{cov}(X_t, X_s) = E[(X_t - \mu_X(t))(X_s - \mu_X(s))], \quad t, s \in T.$$

> The *variance function* of X is given by
>
> $$\sigma_X^2(t) = c_X(t,t) = \text{var}(X_t), \quad t \in T.$$

We learnt in Example 1.2.5 that Gaussian processes are determined only via their expectation and covariance functions. This is not correct for a non-Gaussian process.

As for a random vector, the expectation function $\mu_X(t)$ is a deterministic quantity around which the sample paths of X are concentrated. The covariance function $c_X(t,s)$ is a measure of dependence in the process X. The variance function $\sigma_X^2(t)$ can be considered as a measure of spread of the sample paths of X around $\mu_X(t)$. In contrast to the one-dimensional case, a statement like "95% of all sample paths lie between the graphs of $\mu_X(t) - 2\sigma_X(t)$ and $\mu_X(t) + 2\sigma_X(t)$" is very difficult to show (even for Gaussian processes), and is in general not correct. We will sometimes consider computer graphs with paths of certain stochastic processes and also indicate the curves $\mu_X(t)$ and $\mu_X(t) \pm 2\sigma_X(t)$, $t \in T$. The latter have to be interpreted for every fixed t, i.e. for every individual random variable X_t. Only in a heuristic sense, do they give bounds for the paths of the process X. See Figure 1.2.6 for an illustration.

Example 1.2.7 (Continuation of Example 1.2.5)
Consider the Gaussian process $(X_t, t \in [0,1])$ of iid $N(0,1)$ random variables X_t. Its expectation and covariance functions are given by

$$\mu_X(t) = 0 \quad \text{and} \quad c_X(t,s) = \begin{cases} 1 & \text{if} \quad t = s, \\ 0 & \text{if} \quad t \neq s. \end{cases} \qquad \square$$

Dependence Structure

We have already introduced Gaussian processes by specifying their fidis as multivariate Gaussian. Another way of classifying stochastic processes consists of imposing a special dependence structure.

The process $X = (X_t, t \in T)$, $T \subset \mathbb{R}$, is *strictly stationary* if the fidis are invariant under shifts of the index t:

$$(X_{t_1}, \ldots, X_{t_n}) \stackrel{d}{=} (X_{t_1+h}, \ldots, X_{t_n+h}) \tag{1.9}$$

for all possible choices of indices $t_1, \ldots, t_n \in T$, $n \geq 1$ and h such that $t_1 + h, \ldots, t_n + h \in T$. Here $\stackrel{d}{=}$ stands for the identity of the distributions; see p. 211 for the definition. For the random vectors in (1.9) this means that their distribution functions are identical.

Example 1.2.8 (Stationary Gaussian processes)
Consider a process $X = (X_t, t \in T)$ with $T = [0, \infty)$ or $T = \mathbb{Z}$. A trivial example of a strictly stationary process is a sequence of iid random variables X_t, $t \in \mathbb{Z}$. Since a Gaussian process X is determined by its expectation and covariance functions, condition (1.9) reduces to

$$\mu_X(t+h) = \mu_X(t) \quad \text{and} \quad c_X(t,s) = c_X(t+h, s+h)$$

for all $s, t \in T$ such that $s + h, t + h \in T$. But this means that $\mu_X(t) = \mu_X(0)$ for all t, whereas $c_X(t,s) = \tilde{c}_X(|t-s|)$ for some function $\tilde{c}_X$ of one variable. Hence, for a Gaussian process, strict stationarity means that the expectation function is constant and the covariance function only depends on the distance $|t-s|$. More generally, if a (possibly non-Gaussian) process X has the two aforementioned properties, it is called a *stationary (in the wide sense)* or (*second-order*) *stationary* process.   □

If we describe a real-life process by a (strictly or in the wide sense) stationary stochastic process, then we believe that the characteristic properties of this process do not change when time goes by. The dependence structure described by the fidis or the covariance function is invariant under shifts of time. This is a relatively strong restriction on the underlying process. However, it is a standard assumption in many probability related fields such as statistics and time series analysis.

Stationarity can also be imposed on the increments of a process. The process itself is then not necessarily stationary.

Let $X = (X_t, t \in T)$ be a stochastic process and $T \subset \mathbb{R}$ be an interval. X is said to have *stationary increments* if

$$X_t - X_s \stackrel{d}{=} X_{t+h} - X_{s+h} \text{ for all } t, s \in T \text{ and } h \text{ with } t+h, s+h \in T.$$

X is said to have *independent increments* if for every choice of $t_i \in T$ with $t_1 < \cdots < t_n$ and $n \geq 1$,

$$X_{t_2} - X_{t_1}, \ldots, X_{t_n} - X_{t_{n-1}}$$

are independent random variables.

One of the prime examples of processes with independent, stationary increments is the homogeneous Poisson process. Homogeneity is here another wording for stationarity of the increments.

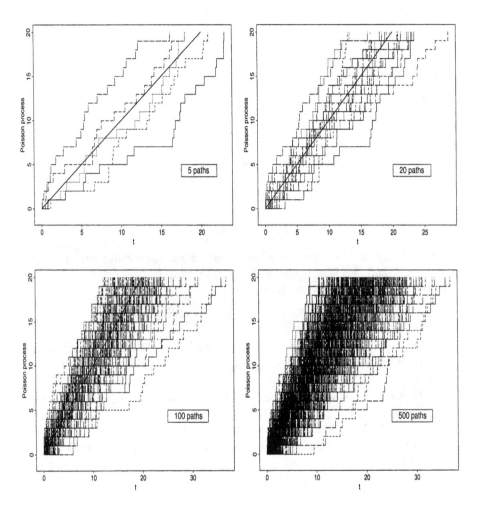

Figure 1.2.9 *Sample paths of a homogeneous Poisson process* $(X_t, t \in [0, \infty))$ *with intensity* $\lambda = 1$; *see Example 1.2.10. The straight solid line stands for the expectation function* $\mu_X(t) = t$.

Example 1.2.10 (Homogeneous Poisson process)

A stochastic process $(X_t, t \in [0, \infty))$ is called an *homogeneous Poisson process* or simply a *Poisson process with intensity* or *rate* $\lambda > 0$ if the following conditions are satisfied:

- It starts at zero: $X_0 = 0$.

- It has stationary, independent increments.

- For every $t > 0$, X_t has a Poisson $Poi(\lambda t)$ distribution; see Example 1.1.1 for the definition of the Poisson distribution.

Figure 1.2.9 shows several Poissonian sample paths.

Notice that, by stationarity of the increments, $X_t - X_s$ with $t > s$ has the same distribution as $X_{t-s} - X_0 = X_{t-s}$, i.e. a $Poi(\lambda(t - s))$ distribution.

An alternative definition of the Poisson process is given by

$$X_t = \#\{n : T_n \le t\}, \quad t > 0, \tag{1.10}$$

where $\#A$ denotes the number of elements of any particular set A, $T_n = Y_1 + \cdots + Y_n$ and (Y_i) is a sequence of iid exponential $Exp(\lambda)$ random variables with common distribution function

$$P(Y_1 \le x) = 1 - e^{-\lambda x}, \quad x \ge 0.$$

This definition shows nicely what kind of sample path a Poisson process has. It is a pure jump function: it is constant on $[T_n, T_{n+1})$ and has upward jumps of size 1 at the random times T_n.

The rôle of the Poisson process and its modifications and ramifications is comparable with the rôle of Brownian motion. The Poisson process is a counting process; see (1.10). It has a large variety of applications in the most different fields. To name a few: for a given time interval $[0, t]$, X_t is a model for the number of

- telephone calls to be handled by an operator,

- customers waiting for service in a queue,

- claims arriving in an insurance portfolio.   □

Notes and Comments

Introductions to the theory of stochastic processes are based on non-elementary facts from measure theory and functional analysis. Standard texts are Ash and Gardner (1975), Gikhman and Skorokhod (1975), Karlin and Taylor (1975,1981) and many others. An entertaining introduction to the theory of applied stochastic processes is Resnick (1992). Grimett and Stirzaker (1994) is an introduction "without burdening the reader with a great deal of measure theory".

1.3 Brownian Motion

1.3.1 Defining Properties

Brownian motion plays a central rôle in probability theory, the theory of stochastic processes, physics, finance,..., and also in this book. We start with the definition of this important process. Then we continue with some of its elementary properties.

A stochastic process $B = (B_t, t \in [0, \infty))$ is called (*standard*) *Brownian motion* or a *Wiener process* if the following conditions are satisfied:

- It starts at zero: $B_0 = 0$.

- It has stationary, independent increments; see p. 30 for the definition.

- For every $t > 0$, B_t has a normal $N(0, t)$ distribution.

- It has continuous sample paths: "no jumps".

See Figure 1.3.1 for a visualization of Brownian sample paths.

Brownian motion is named after the biologist Robert Brown whose research dates to the 1820s. Early in this century, Louis Bachelier (1900), Albert Einstein (1905) and Norbert Wiener (1923) began developing the mathematical theory of Brownian motion. The construction of Bachelier (1900) was erroneous but it captured many of the essential properties of the process. Wiener (1923) was the first to put Brownian motion on a firm mathematical basis.

Distribution, Expectation and Covariance Functions

The fidis of Brownian motion are multivariate Gaussian, hence B is a Gaussian process. Check this statement by observing that Brownian motion has

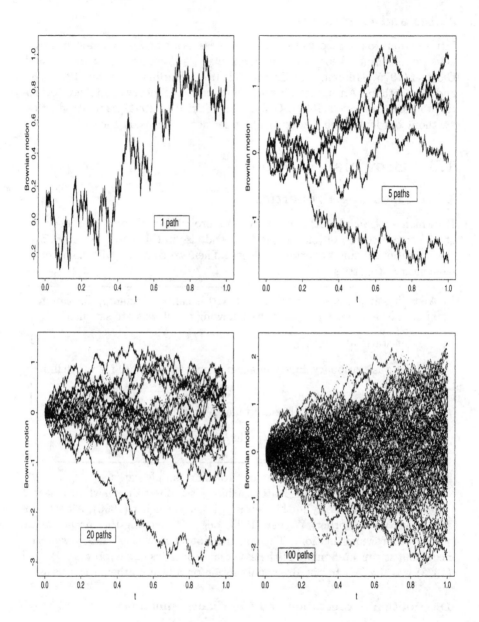

Figure 1.3.1 *Sample paths of Brownian motion on* $[0, 1]$.

independent Gaussian increments and by using the formula for linear transformations of a Gaussian random vector; see p. 18.

> The random variables $B_t - B_s$ and B_{t-s} have an $N(0, t-s)$ distribution for $s < t$.

This follows from the stationarity of the increments. Indeed, $B_t - B_s$ has the same distribution as $B_{t-s} - B_0 = B_{t-s}$ which is normal with mean zero and variance $t - s$. Thus the variance is proportional to the length of the interval $[s, t]$. This means intuitively: the larger the interval, the larger the fluctuations of Brownian motion on this interval. This observation is also supported by simulated Brownian sample paths; see for example Figure 1.3.2.

Notice:

The distributional identity $B_t - B_s \stackrel{d}{=} B_{t-s}$ does not *imply pathwise identity: in general,*

$$B_t(\omega) - B_s(\omega) \neq B_{t-s}(\omega).$$

It is worthwhile to compare Brownian motion with the Poisson process; see Example 1.2.10. Their definitions coincide insofar that they are processes with stationary, independent increments. The crucial difference is the kind of distribution of the increments. The requirement of the Poisson distribution makes the sample paths pure jump functions, whereas the Gaussian assumption makes the sample paths continuous.

It is immediate from the definition that Brownian motion has expectation function

$$\mu_B(t) = EB_t = 0, \quad t \geq 0,$$

and, since the increments $B_s - B_0 = B_s$ and $B_t - B_s$ are independent for $t > s$, it has covariance function (recall its definition from p. 28)

$$
\begin{aligned}
c_B(t, s) &= E\big[[(B_t - B_s) + B_s] B_s\big] = E[(B_t - B_s) B_s] + EB_s^2 \\
&= E(B_t - B_s) EB_s + s = 0 + s = s, \quad 0 \leq s < t.
\end{aligned}
$$

Since a Gaussian process is characterized by its expectation and covariance functions (see Example 1.2.5), we can give an alternative definition:

> Brownian motion is a Gaussian process with
>
> $$\mu_B(t) = 0 \quad \text{and} \quad c_B(t, s) = \min(s, t). \tag{1.11}$$

Path Properties: Non-Differentiability and Unbounded Variation

In what follows, we fix one sample path $B_t(\omega)$, $t \geq 0$, and consider its properties. We already know from the definition of Brownian motion that its sample paths are continuous. However, a glance at simulated Brownian paths immediately convinces us that these functions of t are extremely irregular: they oscillate wildly. The main reason is that the increments of B are independent. In particular, increments of Brownian motion on adjacent intervals are independent whatever the length of the intervals. Since we can imagine the sample path as constructed from its independent increments on adjacent intervals, it is rather surprising that continuity of the path results.

Thus:

How irregular is a Brownian sample path?

Before we answer this question we make a short excursion to a class of stochastic processes which contains Brownian motion as a special case. All members of this class have irregular sample paths.

A stochastic process $(X_t, t \in [0, \infty))$ is *H-self-similar* for some $H > 0$ if its fidis satisfy the condition

$$\left(T^H B_{t_1}, \ldots, T^H B_{t_n}\right) \overset{d}{=} \left(B_{Tt_1}, \ldots, B_{Tt_n}\right) \qquad (1.12)$$

for every $T > 0$, any choice of $t_i \geq 0$, $i = 1, \ldots, n$, and $n \geq 1$.

Notice:

Self-similarity is a distributional, not a pathwise property. In (1.12), one must not replace $\overset{d}{=}$ with $=$.

Roughly speaking, self-similarity means that the properly scaled patterns of a sample path in any small or large time interval have a similar shape, but they are *not* identical. See Figure 1.3.2 for an illustration.

The sample paths of a self-similar process are nowhere differentiable; see Proposition A3.1 on p. 188. And here it comes:

Brownian motion is 0.5-self-similar, i.e.

$$\left(T^{1/2} B_{t_1}, \ldots, T^{1/2} B_{t_n}\right) \overset{d}{=} \left(B_{Tt_1}, \ldots, B_{Tt_n}\right) \qquad (1.13)$$

for every $T > 0$, any choice of $t_i \geq 0$, $i = 1, \ldots, n$, and $n \geq 1$.
Hence its sample paths are nowhere differentiable.

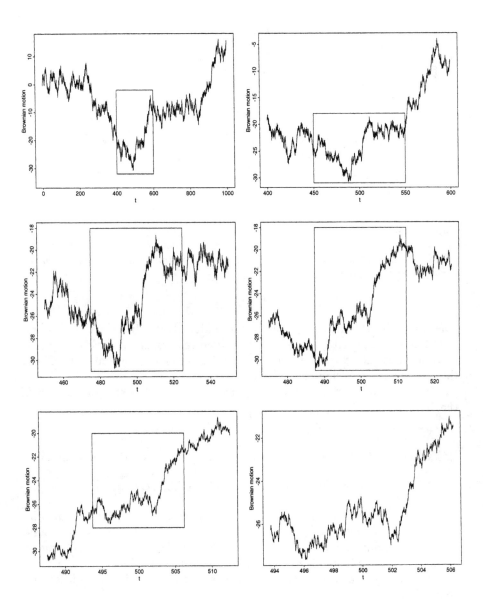

Figure 1.3.2 Self-similarity: *the same Brownian sample path on different scales. The shapes of the curves on different intervals look similar, but they are not simply scaled copies of each other.*

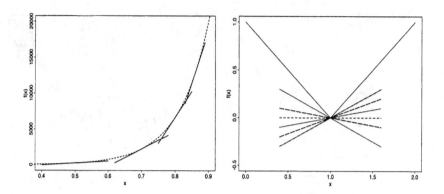

Figure 1.3.3 Left: *a differentiable function. At every point its graph can be approx-*
imated by a linear function which is the unique tangent at this point. Right: *this*
function is not differentiable at x = 1. There are infinitely many tangents to the
curve of the function at this point.

One can easily check the distributional identity (1.13). Indeed, the left- and
right-hand sides of (1.13) are Gaussian random vectors, and therefore it suffices
to verify that they have the same expectation and covariance matrix. Check
these properties by using (1.11).

Differentiability of a function f means that its graph is smooth. Indeed, if
the limit

$$f'(x_0) = \lim_{\Delta x \to 0} \frac{f(x_0 + \Delta x) - f(x_0)}{\Delta x}$$

exists and is finite for some $x_0 \in (0, t)$, say, then we may write for small Δx

$$f(x_0 + \Delta x) = f(x_0) + f'(x_0)\Delta x + h(x_0, \Delta x)\Delta x\,,$$

where $h(x_0, \Delta x) \to 0$ as $\Delta x \to 0$. Hence, in a small neighborhood of x_0, the
function f is roughly linear (as a function of Δx). This explains its smoothness.
Alternatively, differentiability of f at x_0 implies that we have a unique tangent
to the curve of the function f at this point; see Figure 1.3.3 for an illustration.
In this figure you can also see a function which is not differentiable at one
point.

Now try to imagine a nowhere differentiable function: the graph of this
function changes its shape in the neighborhood of any point in a completely
non-predictable way. You will admit that you cannot really imagine such a

function: it is physically impossible. Nevertheless, Brownian motion is considered as a very good approximation to many real-life phenomena. We will see in Section 1.3.3 that Brownian motion is a limit process of certain sum processes.

The self-similarity property of Brownian motion has a nice consequence for the simulation of its sample paths. In order to simulate a path on $[0, T]$ it suffices to simulate one path on $[0, 1]$, then scale the time interval by the factor T and the sample path by the factor $T^{1/2}$. Then we are done.

In some books one can find the claim that the limit

$$\lim_{\Delta t \to 0} \text{var} \left(\frac{B_{t_0 + \Delta t} - B_{t_0}}{\Delta t} \right) \tag{1.14}$$

does not exist and *therefore* the sample paths of Brownian motion are non-differentiable. It is easy to check (do it!) that the limit (1.14) does not exist, but without further theory it would be wrong to conclude from this distributional result that the paths of the process are non-differentiable.

The existence of a nowhere differentiable continuous function was discovered in the 19th century. Such a function was constructed by Weierstrass. It was considered as a curiosity, far away from any practical application. Brownian motion is a process with nowhere differentiable sample paths. Currently it is considered as one of those processes which have a multitude of applications in very different fields. One of them is stochastic calculus; see Chapters 2 and 3.

A further indication of the irregularity of Brownian sample paths is given by the following fact:

Brownian sample paths do not have bounded variation on any finite interval $[0, T]$. This means that

$$\sup_{\tau} \sum_{i=1}^{n} \left| B_{t_i}(\omega) - B_{t_{i-1}}(\omega) \right| = \infty,$$

where the supremum (see p. 211 for its definition) is taken over all possible partitions $\tau : 0 = t_0 < \cdots < t_n = T$ of $[0, T]$.

A proof of this fact is provided by Proposition A3.2 on p. 189. We mention at this point that the unbounded variation and non-differentiability of Brownian sample paths are major reasons for the failure of classical integration methods, when applied to these paths, and for the introduction of stochastic calculus.

1.3.2 Processes Derived from Brownian Motion

The purpose of this section is to get some feeling for the distributional and pathwise properties of Brownian motion. If you want to start with Chapter 2 on stochastic calculus as soon as possible, you can easily skip this section and return to it whenever you need a reference to a property or definition.

Various Gaussian and non-Gaussian stochastic processes of practical relevance can be derived from Brownian motion. Below we introduce some of those processes which will find further applications in the course of this book. As before, $B = (B_t, t \in [0, \infty))$ denotes Brownian motion.

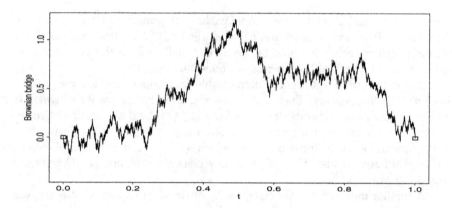

Figure 1.3.4 *A sample path of the Brownian bridge.*

Example 1.3.5 (Brownian bridge)
Consider the process

$$X_t = B_t - t\,B_1\,, \quad 0 \le t \le 1\,.$$

Obviously,

$$X_0 = B_0 - 0\,B_1 = 0 \quad \text{and} \quad X_1 = B_1 - 1\,B_1 = 0\,.$$

For this simple reason, the process X bears the name (standard) *Brownian bridge* or *tied down Brownian motion*. A glance at the sample paths of this "bridge" (see Figure 1.3.4) may or may not convince you that this name is justified.

Using the formula for linear transformations of Gaussian random vectors (see p. 18), one can show that the fidis of X are Gaussian. Verify this! Hence X is a Gaussian process. You can easily calculate the expectation and covariance functions of the Brownian bridge:

$$\mu_X(t) = 0 \quad \text{and} \quad c_X(t,s) = \min(t,s) - ts, \quad s,t \in [0,1].$$

Since X is Gaussian, the Brownian bridge is characterized by these two functions.

The Brownian bridge appears as the limit process of the normalized *empirical distribution function* of a sample of iid uniform $U(0,1)$ random variables. This is a fundamental result from non-parametric statistics; it is the basis for numerous goodness-of-fit tests in statistics. See for example Shorack and Wellner (1986). □

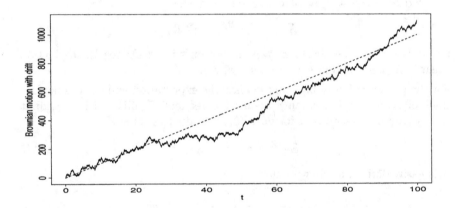

Figure 1.3.6 *A sample path of Brownian motion with drift* $X_t = 20\,B_t + 10\,t$ *on* $[0,100]$. *The dashed line stands for the drift function* $\mu_X(t) = 10\,t$.

Example 1.3.7 (Brownian motion with drift)
Consider the process
$$X_t = \mu\,t + \sigma\,B_t, \quad t \geq 0,$$

for constants $\sigma > 0$ and $\mu \in \mathbb{R}$. Clearly, it is a Gaussian process (why?) with expectation and covariance functions

$$\mu_X(t) = \mu\,t \quad \text{and} \quad c_X(t,s) = \sigma^2 \min(t,s), \quad s,t \geq 0.$$

The expectation function $\mu_X(t) = \mu t$ (the deterministic "drift" of the process) essentially determines the characteristic shape of the sample paths; see Figure 1.3.6 for an illustration. Therefore X is called *Brownian motion with (linear) drift.*   □

With the fundamental discovery of Bachelier in 1900 that prices of risky assets (stock indices, exchange rates, share prices, etc.) can be well described by Brownian motion, a new area of applications of stochastic processes was born. However, Brownian motion, as a Gaussian process, may assume negative values, which is not a very desirable property of a price. In their celebrated papers from 1973, Black, Scholes and Merton suggested another stochastic process as a model for speculative prices. In Section 4.1 we consider their approach to the pricing of European call options in more detail. It is one of the promising and motivating examples for the use of stochastic calculus.

Example 1.3.8 (Geometric Brownian motion)
The process suggested by Black, Scholes and Merton is given by

$$X_t = e^{\mu t + \sigma B_t}, \quad t \geq 0,$$

i.e. it is the exponential of Brownian motion with drift; see Example 1.3.7. Clearly, X is not a Gaussian process (why?).

For the purpose of later use, we calculate the expectation and covariance functions of geometric Brownian motion. For readers, familiar with probability theory, you may recall that for an $N(0,1)$ random variable Z,

$$Ee^{\lambda Z} = e^{\lambda^2/2}, \quad \lambda \in \mathbb{R}. \tag{1.15}$$

It is easily derived as shown below:

$$
\begin{aligned}
Ee^{\lambda Z} &= \frac{1}{(2\pi)^{1/2}} \int_{-\infty}^{\infty} e^{\lambda z} e^{-z^2/2}\, dz \\
&= e^{\lambda^2/2} \frac{1}{(2\pi)^{1/2}} \int_{-\infty}^{\infty} e^{-(z-\lambda)^2/2}\, dz \\
&= e^{\lambda^2/2}.
\end{aligned}
$$

Here we used the fact that $(2\pi)^{-1/2} \exp\{-(z-\lambda)^2/2\}$ is the density of an $N(\lambda, 1)$ random variable.

From (1.15) and the self-similarity of Brownian motion it follows immediately that

$$\mu_X(t) = e^{\mu t} Ee^{\sigma B_t} = e^{\mu t} Ee^{\sigma t^{1/2} B_1} = e^{(\mu + 0.5\sigma^2)t}. \tag{1.16}$$

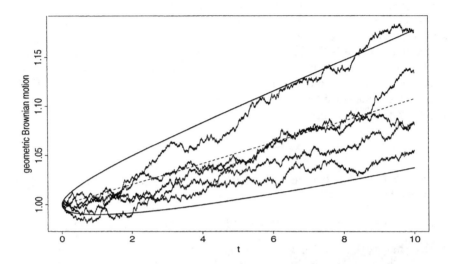

Figure 1.3.9 *Sample paths of geometric Brownian motion $X_t = \exp\{0.01t + 0.01B_t\}$ on $[0, 10]$, the expectation function $\mu_X(t)$ (dashed line) and the graphs of the functions $\mu_X(t) \pm 2\sigma_X(t)$ (solid lines). The latter curves have to be interpreted with care since the distributions of the $X_t s$ are not normal.*

For $s \leq t$, $B_t - B_s$ and B_s are independent, and $B_t - B_s \overset{d}{=} B_{t-s}$. Hence

$$
\begin{aligned}
c_X(t, s) &= EX_t X_s - EX_t EX_s \qquad\qquad (1.17)\\
&= e^{\mu(t+s)} E e^{\sigma(B_t + B_s)} - e^{(\mu + 0.5\,\sigma^2)(t+s)}\\
&= e^{\mu(t+s)} E e^{\sigma[(B_t - B_s) + 2B_s]} - e^{(\mu + 0.5\,\sigma^2)(t+s)}\\
&= e^{\mu(t+s)} E e^{\sigma(B_t - B_s)} E e^{2\sigma B_s} - e^{(\mu + 0.5\,\sigma^2)(t+s)}\\
&= e^{(\mu + 0.5\,\sigma^2)(t+s)} \left(e^{\sigma^2 s} - 1 \right).
\end{aligned}
$$

In particular, geometric Brownian motion has variance function

$$
\sigma_X^2(t) = e^{(2\mu + \sigma^2)t} \left(e^{\sigma^2 t} - 1 \right). \qquad\qquad (1.18)
$$

See Figure 1.3.9 for an illustration of various sample paths of geometric Brownian motion. $\qquad\qquad\qquad\qquad\qquad\qquad\qquad\qquad\qquad\qquad\qquad\square$

Example 1.3.10 (Gaussian white and colored noise)
In statistics and time series analysis one often uses the name "white noise" for a sequence of iid or uncorrelated random variables. This is in contrast to physics, where white noise is understood as a certain derivative of Brownian sample paths. This does not contradict our previous remarks since this derivative is not obtained by ordinary differentiation. Since white noise is "physically impossible", one considers an approximation to it, called *colored noise*. It is a Gaussian process defined as

$$X_t = \frac{B_{t+h} - B_t}{h}, \quad t \geq 0, \tag{1.19}$$

where $h > 0$ is some fixed constant. Its expectation and covariance functions are given by

$$\mu_X(t) = 0 \quad \text{and} \quad c_X(t,s) = h^{-2}[(s+h) - \min(s+h,t)], \quad s \leq t.$$

Notice that $c_X(t,s) = 0$ if $t - s \geq h$, hence X_t and X_s are independent, but if $t - s < h$, $c_X(t,s) = h^{-2}[h - (t-s)]$. Since X is Gaussian and $c_X(t,s)$ is a function only of $t - s$, it is stationary (see Example 1.2.8).

Clearly, if B was differentiable, we could let h in (1.19) go to zero, and in the limit we would obtain the ordinary derivative of B at t. But, as we know, this argument is not applicable. The variance function $\sigma_X^2(t) = h^{-1}$ gives an indication that the fluctuations of colored noise become larger as h decreases. Simulated paths of colored noise look very much like the sample paths in Figure 1.2.6.   □

1.3.3   Simulation of Brownian Sample Paths

This section is not necessary for the understanding of stochastic calculus. However, it will characterize Brownian motion as a distributional limit of partial sum processes (so-called functional central limit theorem). This observation will help you to understand the Brownian path properties (non-differentiability, unbounded variation) much better. A second objective of this section is to show that Brownian sample paths can easily be simulated by using standard software.

Using the almost unlimited power of modern computers, you can visualize the paths of almost every stochastic process. This is desirable because we like to see sample paths in order to understand the stochastic process better. On the other hand, simulations of the paths of stochastic processes are sometimes unavoidable if you want to say something about the distributional properties of such a process. In most cases, we cannot determine the exact distribution of a stochastic process and its functionals (such as its maximum or minimum on a given interval). Then simulations and numerical techniques offer some alternative to calculate these distributions.

Simulation via the Functional Central Limit Theorem

From an elementary course in probability theory we know about the *central limit theorem (CLT)*. It is a fundamental result: it explains why the normal distribution plays such an important rôle in probability theory and statistics. The CLT says that the properly normalized and centered partial sums of an iid finite variance sequence converge in distribution to a normal distribution. To be precise: let $Y_1, Y_2, \ldots$ be iid non-degenerate (i.e. non-constant) random variables with mean $\mu_Y = EY_1$ and variance $\sigma_Y^2 = \mathrm{var}(Y_1)$. Define the partial sums

$$R_0 = 0, \quad R_n = Y_1 + \cdots + Y_n, \quad n \geq 1.$$

Recall that Φ denotes the distribution function of a standard normal random variable.

If Y_1 has finite variance, then the sequence (Y_i) obeys the CLT, i.e.

$$\sup_x \left| P\left(\frac{R_n - ER_n}{[\mathrm{var}(R_n)]^{1/2}} \leq x \right) - \Phi(x) \right| \to 0 \quad \text{as} \quad n \to \infty.$$

Thus, for large n, the distribution of $(R_n - \mu_Y n)/(\sigma_Y^2 n)^{1/2}$ is approximately standard normal. This is an amazing fact, since the CLT holds independently of the distribution of the Y_is; all one needs is a finite variance σ_Y^2.

The CLT has an analogue for stochastic processes. Consider the process with continuous sample paths on $[0, 1]$:

$$S_n(t) = \begin{cases} (\sigma_Y^2 n)^{-1/2} (R_i - \mu_Y i), & \text{if } t = i/n, \, i = 0, \ldots, n, \\ \text{linearly interpolated}, & \text{elsewhere.} \end{cases} \tag{1.20}$$

In Figures 1.3.11 and 1.3.12 you will find realizations of S_n for various n.

Assume for the moment that the Y_is are iid $N(0, 1)$ and consider the restriction of the process S_n to the points i/n. We immediately see that the following properties hold:

- S_n starts at zero: $S_n(0) = 0$.

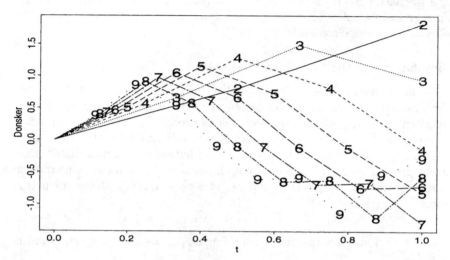

Figure 1.3.11 *Sample paths of the process S_n for one sequence of realizations $Y_1(\omega), \ldots, Y_9(\omega)$ and $n = 2, \ldots, 9$.*

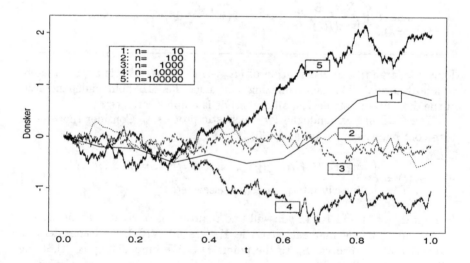

Figure 1.3.12 *Sample paths of the process S_n for different n and the same sequence of realizations $Y_1(\omega), \ldots, Y_{100,000}(\omega)$.*

- S_n has independent increments, i.e. for all integers $0 \leq i_1 \leq \cdots \leq i_m \leq n$, the random variables

$$S_n(i_2/n) - S_n(i_1/n), \ldots, S_n(i_m/n) - S_n(i_{m-1}/n)$$

  are independent.

- For every $0 \leq i \leq n$, $S_n(i/n)$ has a normal $N(0, i/n)$ distribution.

Thus, S_n and Brownian motion B on $[0, 1]$, when restricted to the points i/n, have very much the same properties; cf. the definition of Brownian motion on p. 33. Naturally, the third property above is not valid if we drop the assumption that the Y_is are iid Gaussian. However, in an asymptotic sense the stochastic process S_n is close to Brownian motion:

> If Y_1 has finite variance, then the sequence (Y_i) obeys the *functional CLT*, also called *Donsker's invariance principle*, i.e. the processes S_n converge in distribution to Brownian motion B on $[0, 1]$.

Convergence in distribution of S_n has a two-fold meaning. The first one is quite intuitive:

- The fidis of S_n converge to the corresponding fidis of B, i.e.

$$P\left(S_n(t_1) \leq x_1, \ldots, S_n(t_m) \leq x_m\right) \to P\left(B_{t_1} \leq x_1, \ldots, B_{t_m} \leq x_m\right)$$
$$(1.21)$$

  for all possible choices of $t_i \in [0, 1]$, $x_i \in \mathbb{R}$, $i = 1, \ldots, m$, and all integers $m \geq 1$.

But convergence of the fidis is not sufficient for the convergence in distribution of stochastic processes. Fidi convergence determines the Gaussian limit distribution for every choice of finitely many fixed instants of time t_i, but stochastic processes are infinite-dimensional objects, and therefore unexpected events may happen. For example, the sample paths of the converging processes may fluctuate very wildly with increasing n, in contrast to the limiting process of Brownian motion which has continuous sample paths. In order to avoid such irregular behavior,

- a so-called *tightness* or *stochastic compactness* condition must be satisfied.

Fortunately, the partial sum processes S_n are tight, but it is beyond the scope of this book to show it.

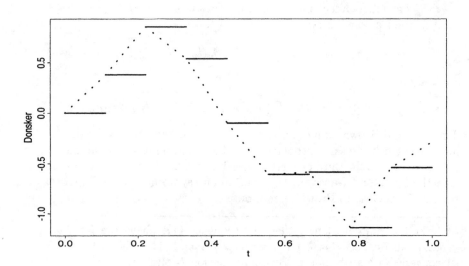

Figure 1.3.13 *One sample path of the processes S_9 (dotted line) and $\widetilde{S}_9$ (solid line) for the same sequence of realizations $Y_1(\omega), \ldots, Y_9(\omega)$. See (1.20) and (1.22) for the definitions of S_n and $\widetilde{S}_n$.*

What was said about the convergence of the processes S_n remains valid for the processes

$$\widetilde{S}_n(t) = (\sigma_Y^2 n)^{-1/2}(R_{[nt]} - \mu_Y\,[nt]), \quad 0 \le t \le 1, \tag{1.22}$$

where $[nt]$ denotes the integer part of the real number nt; see Figure 1.3.13 for an illustration. In contrast to S_n, the process $\widetilde{S}_n$ is constant on the intervals $[(i-1)/n, i/n)$ and has jumps at the points i/n. But S_n and $\widetilde{S}_n$ coincide at the points i/n, and the differences between these two processes are asymptotically negligible: the normalization $n^{1/2}$ makes the jumps of $\widetilde{S}_n$ arbitrarily small for large n. As for S_n, we can formulate the following functional CLT:

> If Y_1 has finite variance, then the sequence (Y_i) obeys the *functional CLT*, i.e. the processes $\widetilde{S}_n$ converge in distribution to Brownian motion B on $[0, 1]$.

Since $\widetilde{S}_n$ is a jump process, the notion of convergence in distribution becomes even more complicated than for S_n. We refrain from discussing details.

Thus we have a first simple tool for simulating Brownian sample paths in our hands:

Plot the paths of the processes S_n or $\widetilde{S}_n$ for sufficiently large n, and you get a reasonable approximation to Brownian sample paths.

Also notice:

Since Brownian motion appears as a distributional limit you will see completely different graphs for different values of n, for the same sequence of realizations $Y_i(\omega)$.

From the self-similarity property we also know how to obtain an approximation to the sample paths of Brownian motion on any interval $[0, T]$:

Simulate one path of S_n or $\widetilde{S}_n$ on $[0, 1]$, then scale the time interval by the factor T and the sample path by the factor $T^{1/2}$.

Standard software (such as Splus, Mathematica, Matlab, etc.) provides you with quick and reliable algorithms for generating random numbers of standard distributions. Random number generators are frequently based on natural processes, for example radiation, or on algebraic methods. The generated "random" numbers can be considered as "pseudo" realizations $Y_i(\omega)$ of iid random variables Y_i.

For practical purposes, you may want to choose the realizations $Y_i(\omega)$ (or, as you like, the random numbers $Y_i(\omega)$) from an appropriate distribution. If you are interested in "good" approximations to the Gaussian distribution of Brownian motion, you would generate the $Y_i(\omega)$s from a Gaussian distribution. If you are forced to simulate many sample paths of S_n or $\widetilde{S}_n$ in a short period of time, you would perhaps choose the $Y_i(\omega)$s as realizations of iid Bernoulli random variables, i.e. $P(Y_i = \pm 1) = 0.5$, or of iid uniform random variables.

Simulation via Series Representations

Recall from a course on calculus that every continuous 2π-periodic function f on $\mathbb{R}$ (i.e. $f(x + 2\pi) = f(x)$ for all $x \in \mathbb{R}$) has a Fourier series representation, i.e. it can be written as an infinite series of trigonometric functions.

Since Brownian sample paths are continuous functions, we can try to expand them in a Fourier series. However, the paths are random functions: for different ω we obtain different functions. This means that the coefficients of this Fourier series are random variables, and since the process is Gaussian, they must be Gaussian as well.

The following representation of Brownian motion on the interval $[0, 2\pi]$ is called *Paley–Wiener representation*: let $(Z_n, n \geq 0)$ be a sequence of iid

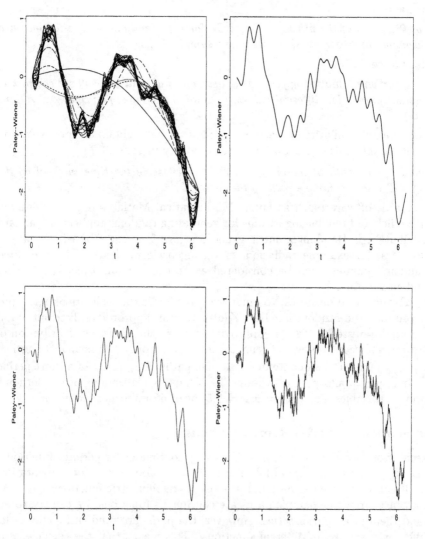

Figure 1.3.14 *Simulation of one Brownian sample path from the discretization* (1.24) *of the Paley–Wiener representation with* $N = 1,000$. Top left: *all paths for* $M = 2, \ldots, 40$. Top right: *the path only for* $M = 40$. Bottom left: $M = 100$. Bottom right: $M = 800$.

$N(0,1)$ random variables, then

$$B_t(\omega) = Z_0(\omega)\frac{t}{(2\pi)^{1/2}} + \frac{2}{\pi^{1/2}} \sum_{n=1}^{\infty} Z_n(\omega)\frac{\sin(nt/2)}{n}, \quad t \in [0, 2\pi]. \quad (1.23)$$

This series converges for every fixed t, but also uniformly for $t \in [0, 2\pi]$, i.e. the rate of convergence is comparable for all t. For an application of this formula, one has to decide about the number M of sine functions and the number N of discretization points at which the sine functions will be evaluated. This amounts to calculating the values

$$Z_0(\omega)\frac{t_j}{(2\pi)^{1/2}} + \frac{2}{\pi^{1/2}} \sum_{n=1}^{M} Z_n(\omega)\frac{\sin(nt_j/2)}{n}, \quad (1.24)$$

$$t_j = \frac{2\pi j}{N}, \quad j = 0, 1, \dots, N.$$

The problem of choosing the "right" values for M and N is similar to the choice of the sample size n in the functional CLT; it is difficult to give a simple rule of thumb for the choices of M and N.

In Figure 1.3.14 you can see that the shape of the sample paths does not change very much if one switches from $M = 100$ to $M = 800$ sine functions. A visual inspection in the case $M = 100$ gives the impression that the sample path is still too smooth. This is not completely surprising since a sum of M sine functions is differentiable; only in the limit (as $M \to \infty$) do we gain a non-differentiable sample path.

The Paley–Wiener representation is just one of infinitely many possible series representations of Brownian motion. Another well-known such representation is due to Lévy. In the *Lévy representation*, the sine functions are replaced by certain polygonal functions (the Schauder functions).

To be precise, first define the *Haar functions* H_n on $[0, 1]$ as follows:

$$H_1(t) = 1,$$

$$H_{2^m+1}(t) = \begin{cases} 2^{m/2}, & \text{if } t \in \left[1 - \dfrac{2}{2^{m+1}}, 1 - \dfrac{1}{2^{m+1}}\right), \\[2mm] -2^{m/2}, & \text{if } t \in \left[1 - \dfrac{1}{2^{m+1}}, 1\right], \\[2mm] 0, & \text{elsewhere}, \end{cases}$$

$$H_{2^m+k}(t) = \begin{cases} 2^{m/2}, & \text{if } t \in \left[\dfrac{k-1}{2^m}, \dfrac{k-1}{2^m} + \dfrac{1}{2^{m+1}}\right), \\[2ex] -2^{m/2}, & \text{if } t \in \left[\dfrac{2k-1}{2^{m+1}}, \dfrac{k}{2^m}\right), \\[2ex] 0, & \text{elsewhere}, \end{cases}$$

$$k = 1, \ldots, 2^m - 1; \quad m = 0, 1, \ldots.$$

From these functions define the system of the *Schauder functions* on $[0,1]$ by integrating the Haar functions:

$$\tilde{H}_n(t) = \int_0^t H_n(s)\, ds, \quad n = 1, 2, \ldots.$$

Figures 1.3.15 and 1.3.16 show the graphs of H_n and $\tilde{H}_n$ for the first n. A series representation for a Brownian sample path on $[0,1]$ is then given by

$$B_t(\omega) = \sum_{n=1}^{\infty} Z_n(\omega)\tilde{H}_n(t), \quad t \in [0,1], \tag{1.25}$$

where the convergence of this series is uniform for $t \in [0,1]$ and the $Z_n(\omega)$s are realizations of an iid $N(0,1)$ sequence (Z_n). As for simulations of Brownian motion via sine functions, one has to choose a truncation point M of the infinite series (1.25). In Figure 1.3.17 we show how a Brownian sample path is approximated by the superposition of the first M terms in the series representation (1.25). In contrast to Figure 1.3.14, the polygonal shape of the Schauder functions already anticipates the irregular behavior of a Brownian path (its non-differentiability) for relatively small M.

The Paley–Wiener and Lévy representations are just two of infinitely many possible series representations of Brownian motion. They are special cases of the so-called *Lévy–Ciesielski representation*. Ciesielski showed that Brownian motion on $[0,1]$ can be represented in the form

$$B_t(\omega) = \sum_{n=1}^{\infty} Z_n(\omega) \int_0^t \phi_n(x)\,dx, \quad t \in [0,1],$$

where Z_n are iid $N(0,1)$ random variables and (ϕ_n) is a complete orthonormal function system on $[0,1]$.

Notes and Comments

Brownian motion is the best studied stochastic process. Various books are devoted to it, for example Borodin and Salminen (1996), Hida (1980), Karatzas

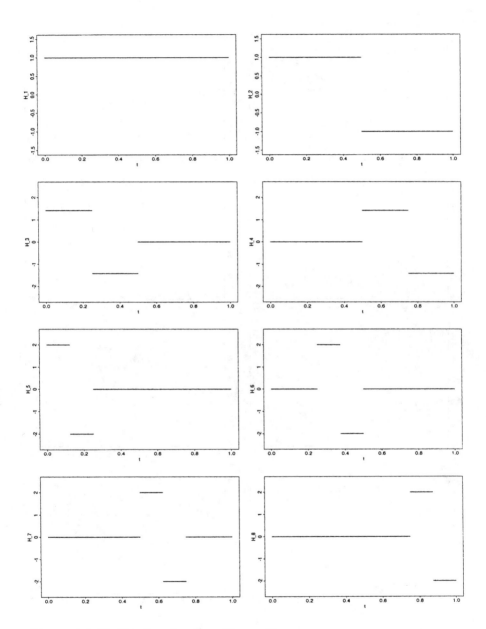

Figure 1.3.15 *The Haar functions $H_1, \ldots, H_8$.*

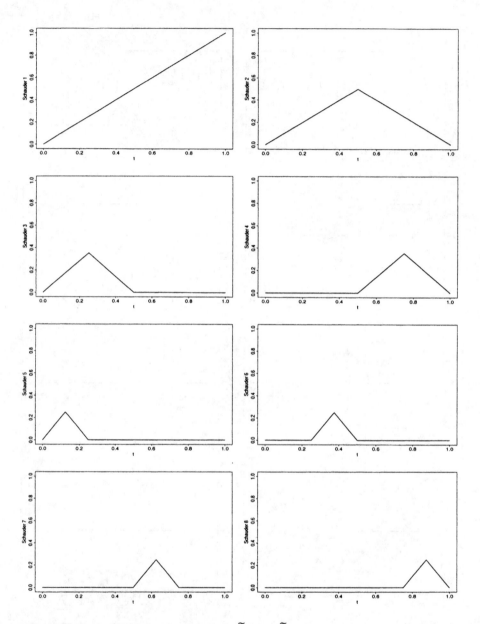

Figure 1.3.16 *The Schauder functions* $\widetilde{H}_1, \ldots, \widetilde{H}_8$.

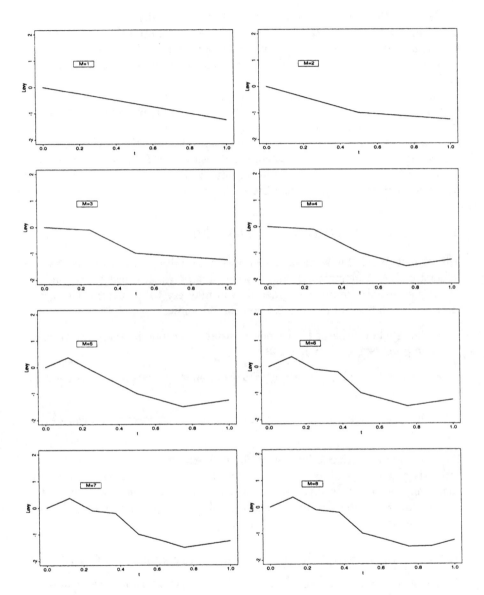

Figure 1.3.17 *The first steps in the construction of one Brownian sample path from the Lévy representation (1.25) via M Schauder functions, M = 1, ..., 8.*

and Shreve (1988) and Revuz and Yor (1991). The reader of these books must be familiar with the theory of stochastic processes, functional analysis, special functions and measure theory. Every textbook on stochastic processes also contains at least one chapter about Brownian motion; see the references on p. 33.

In addition to the non-differentiability and unbounded variation, Brownian motion has many more exciting path and distributional properties. Hida (1980) is a good reference to read about them.

The functional CLT is to be found in advanced textbooks on stochastic processes and the convergence of probability measures; see for example Billingsley (1968) or Pollard (1984). The series representations of Brownian motion can be found in Hida (1980); see also Ciesielski (1965).

1.4   Conditional Expectation

You cannot avoid this section; it contains material which is essential for the understanding of martingales, and more generally, Itô stochastic integrals.

If you are not interested in details you may try to read from one box to another. At the end of Section 1.4 you should know:

- the σ-field generated by a random variable, a random vector or a stochastic process; see Section 1.4.2,

- the conditional expectation of a random variable given a σ-field; see Section 1.4.3,

- the most common rules for calculating conditional expectations; see Section 1.4.4.

You should start with Section 1.4.1, where an example of a conditional expectation is given. It will give you some motivation for the abstract notion of conditional expectation given a σ-field, and every time when you get lost in this section, you should return to Section 1.4.1 and try to figure out what the general theory says in this concrete case.

1.4.1   Conditional Expectation under Discrete Condition

From an elementary course on probability theory we know the *conditional probability of A given B*, i.e.

$$P(A\,|\,B) = \frac{P(A \cap B)}{P(B)}\,.$$

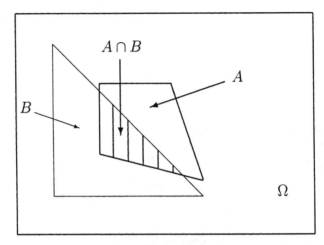

Figure 1.4.1 *The classical conditional probability: if we know that B occurred, we assign the new probability 1 to it. Events outside B cannot occur, hence they have the new probability 0.*

Clearly,

$$P(A \,|\, B) = P(A) \quad \text{if and only if } A \text{ and } B \text{ are independent.}$$

For the definition of $P(A \,|\, B)$ it is crucial that $P(B)$ is positive. It is the objective of Section 1.4.3 to relax this condition.

The probability $P(A \,|\, B)$ can be interpreted as follows. Assume the event B occurred. This is additional information which substantially changes the underlying probability measure. In particular, we assign the new probabilities 0 to B^c (we know that B^c will not happen) and 1 to B. The event B becomes our new probability space Ω', say. All events of interest are now subsets of Ω': $A \cap B \subset \Omega'$. In order to get a new probability measure on Ω' we have to normalize the old probabilities $P(A \cap B)$ by $P(B)$. In sum, the occurrence of B makes our original space Ω shrink to Ω', and the original probabilities $P(A)$ have to be replaced with $P(A \,|\, B)$.

Given that $P(B) > 0$, we can define the *conditional distribution function of a random variable X given B*

$$F_X(x \,|\, B) = \frac{P(X \leq x \,,\, B)}{P(B)}, \quad x \in \mathbb{R},$$

and also the *conditional expectation of X given B*

$$E(X \mid B) = \frac{E(XI_B)}{P(B)}, \tag{1.26}$$

where

$$I_B(\omega) = \begin{cases} 1 & \text{if } \omega \in B, \\ 0 & \text{if } \omega \notin B, \end{cases}$$

denotes the *indicator function of the event B*. In order to discuss the definition (1.26), we assume for the moment that $\Omega = \mathbb{R}$. If X is a discrete random variable with values $x_1, x_2, \ldots$, then (1.26) becomes

$$E(X \mid B) = \sum_{k=1}^{\infty} x_k \frac{P(\{\omega : X(\omega) = x_k\} \cap B)}{P(B)}$$

$$= \sum_{k=1}^{\infty} x_k P(X = x_k \mid B).$$

If X has density f_X, then (1.26) becomes

$$E(X \mid B) = \frac{1}{P(B)} \int_{-\infty}^{\infty} x I_B(x) f_X(x) \, dx$$

$$=: \frac{1}{P(B)} \int_B x f_X(x) \, dx.$$

It is common usage to write $\int_B g(x) dx$ for $\int_{-\infty}^{\infty} g(x) I_B(x) dx$.

Example 1.4.2 (The conditional expectation of a uniform random variable) We consider the random variable $X(\omega) = \omega$ on the space $\Omega = (0, 1]$, endowed with the probability measure P such that

$$P((a, b]) = b - a, \quad (a, b] \subset (0, 1].$$

Clearly, X has a uniform distribution on $(0, 1]$: its distribution function is given by

$$F_X(x) = P(\{\omega : X(\omega) = \omega \le x\})$$

$$= \begin{cases} P(\emptyset) & = 0 & \text{if } x \le 0, \\ P((0, x]) & = x & \text{if } x \in (0, 1], \\ P((0, 1]) & = 1 & \text{if } x > 1. \end{cases}$$

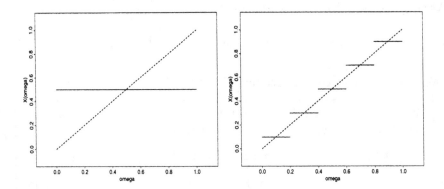

Figure 1.4.3 Left: *a uniform random variable X on $(0, 1]$ (dotted line) and its expectation (solid line); see Example 1.4.2. Right: the random variable X (dotted line) and the conditional expectations $E(X|A_i)$ (solid lines), where $A_i = ((i-1)/5, i/5]$, $i = 1, \ldots, 5$. These conditional expectations can be interpreted as the values of a discrete random variable $E(X|Y)$ with distinct constant values on the sets A_i; see Example 1.4.4.*

Both, the random variable X and its expectation $EX = 0.5$, are represented in Figure 1.4.3.

Now assume that one of the events

$$A_i = ((i-1)/n, i/n], \quad i = 1, \ldots, n,$$

occurred. Recall that $f_X(x) = 1$ on $(0, 1]$ and notice that $P(A_i) = 1/n$. Then

$$E(X \mid A_i) = \frac{1}{P(A_i)} \int_{A_i} x f_X(x) \, dx = n \int_{(i-1)/n}^{i/n} x \, dx = \frac{1}{2} \frac{2i - 1}{n}. \qquad (1.27)$$

The conditional expectations $E(X \mid A_i)$ are illustrated in Figure 1.4.3. The value $E(X \mid A_i)$ is the updated expectation on the new space A_i, given the information that A_i occurred. □

Now we consider a discrete random variable Y on Ω that assumes the distinct values y_i on the sets A_i, i.e.

$$A_i = \{\omega : Y(\omega) = y_i\}, \quad i = 1, 2, \ldots.$$

Clearly, (A_i) is a disjoint partition of Ω, i.e.

$$A_i \cap A_j = \emptyset \quad \text{for} \quad i \neq j \quad \text{and} \quad \bigcup_{i=1}^{\infty} A_i = \Omega. \tag{1.28}$$

We also assume for convenience that $P(A_i) > 0$ for all i.

For a random variable X on Ω with $E|X| < \infty$ we define the *conditional expectation of* X *given* Y as the discrete random variable

$$E(X \mid Y)(\omega) \;\; = \;\; E(X \mid A_i) = E(X \mid Y = y_i) \quad \text{for} \; \omega \in A_i,$$

$$i = 1, 2, \dots. \tag{1.29}$$

If we know that A_i occurred, we may restrict ourselves to ωs in A_i. For those ωs, $E(X \mid Y)(\omega)$ coincides with the classical conditional expectation $E(X \mid A_i)$; see (1.26).

Example 1.4.4 (Continuation of Example 1.4.2)
We interpret the $E(X \mid A_i)$s in (1.27) as the values of a discrete random variable $E(X \mid Y)$, where Y is constant on the sets $A_i = ((i-1)/n, i/n]$. See Figure 1.4.3 for an illustration of this random variable. In this sense, $E(X \mid Y)$ is nothing but a coarser version of the original random variable X, i.e. an approximation to X, given the information that any of the A_is occurred.   □

In what follows we give some elementary properties of the random variable $E(X \mid Y)$. The first property can be easily checked. (Do it!)

The conditional expectation is *linear* : for random variables X_1, X_2 and constants c_1, c_2,

$$E(\,[c_1 \, X_1 \; + \; c_2 \, X_2] \mid Y) = c_1 \, E(X_1 \mid Y) \; + \; c_2 \, E(X_2 \mid Y).$$

The expectations of X and $E(X \mid Y)$ are the same: $EX = E[E(X \mid Y)]$.

This follows by a direct application of the defining properties (1.29), (1.26) and by the observation that $E(X \mid Y)$ is a discrete random variable:

$$E(E(X \mid Y)) \;\; = \;\; \sum_{i=1}^{\infty} E(X \mid A_i) \, P(A_i) = \sum_{i=1}^{\infty} E(X I_{A_i})$$

$$= E\left(X \sum_{i=1}^{\infty} I_{A_i}\right) = EX.$$

Here we also used (1.28) so that

$$\sum_{i=1}^{\infty} I_{A_i} = I_{\bigcup_{i=1}^{\infty} A_i} = I_{\Omega} = 1.$$

If X and Y are independent, then $E(X \mid Y) = EX$. \qquad (1.30)

Recall from p. 20 that independence of X and Y implies

$$P(X \in A, Y = y_i) = P(X \in A) P(Y = y_i) = P(X \in A) P(A_i). \qquad (1.31)$$

Consider the random variable I_{A_i} and notice that

$$\{\omega : I_{A_i}(\omega) = 1\} = A_i = \{\omega : Y(\omega) = y_i\}.$$

Thus we can rewrite (1.31) as follows

$$P(X \in A, I_{A_i} = 1) = P(X \in A) P(I_{A_i} = 1).$$

The analogous relation, where $\{I_{A_i} = 1\}$ is replaced with $\{I_{A_i} = 0\}$, also holds. Hence the random variables X and I_{A_i} are independent and for $\omega \in A_i$,

$$E(X \mid Y)(\omega) = E(X \mid A_i) = \frac{E(X I_{A_i})}{P(A_i)} = \frac{EX\, E(I_{A_i})}{P(A_i)} = EX,$$

where we used that

$$EI_{A_i} = 0\, P(A_i^c) + 1\, P(A_i) = P(A_i).$$

This proves (1.30).

Up to this point, we have learnt:

- The conditional expectation $E(X \mid Y)$ of X given a discrete random variable Y is a discrete random variable.

- It coincides with the classical conditional expectation $E(X \mid Y = y_i)$ on the sets $A_i = \{\omega : Y(\omega) = y_i\}$.

- In this sense, it is a coarser version of X; see again Figure 1.4.3.

- The fewer values Y has, the coarser the random variable $E(X \mid Y)$. In particular, if $Y = const$, then $E(X \mid Y) = EX$; if Y assumes two distinct values, so does $E(X \mid Y)$, etc.

- The conditional expectation $E(X \mid Y)$ is not a function of X, but merely a function of Y. The random variable X only determines the kind of function. Indeed, we can write

$$E(X \mid Y) = g(Y), \quad \text{where} \quad g(y) = \sum_{i=1}^{\infty} E(X \mid Y = y_i) \, I_{\{y_i\}}(y).$$

1.4.2   About σ-Fields

In the previous section we introduced the conditional expectation $E(X \mid Y)$ of a random variable X under the discrete condition (i.e. discrete random variable) Y. Recall from (1.29) that the values of Y did not really matter for the definition of $E(X \mid Y)$, but it was crucial that Y assumed the distinct values y_i on the ω-sets A_i. Thus the conditional expectation $E(X \mid Y)$ can actually be understood as a random variable constructed from a collection $\sigma(Y)$, say, of subsets of Ω. So we may, in a symbolic way, write

$$E(X \mid Y) = E(X \mid \sigma(Y)).$$

Obviously, the collection $\sigma(Y)$ provides us with the information about the structure of the random variable $Y(\omega)$ as a function of $\omega \in \Omega$.

In what follows, we want to make precise what "collection $\sigma(Y)$ of subsets of Ω" means. We will call it a *σ-field* or a *σ-algebra*. Its definition follows:

A *σ-field* $\mathcal{F}$ (on Ω) is a collection of subsets of Ω satisfying the following conditions:

- It is not empty: $\emptyset \in \mathcal{F}$ and $\Omega \in \mathcal{F}$.

- If $A \in \mathcal{F}$, then $A^c \in \mathcal{F}$.

- If $A_1, A_2, \ldots \in \mathcal{F}$, then

$$\bigcup_{i=1}^{\infty} A_i \in \mathcal{F} \quad \text{and} \quad \bigcap_{i=1}^{\infty} A_i \in \mathcal{F}.$$

Example 1.4.5 (Some elementary σ-fields)
Check that the following collections of subsets of Ω are σ-fields:

$$
\begin{aligned}
\mathcal{F}_1 &= \{\emptyset, \Omega\}, \\
\mathcal{F}_2 &= \{\emptyset, \Omega, A, A^c\} \quad \text{for some } A \neq \emptyset \text{ and } A \neq \Omega, \\
\mathcal{F}_3 &= \mathcal{P}(\Omega) = \{A : A \subset \Omega\}.
\end{aligned}
$$

$\mathcal{F}_1$ is the smallest σ-field on Ω, and $\mathcal{F}_3$, the *power set of* Ω, is the biggest one, as it contains all possible subsets of Ω. □

Now suppose that $\mathcal{C}$ is a collection of subsets of Ω, but not necessarily a σ-field. By adding more sets to $\mathcal{C}$, one can always obtain a σ-field, for example the power set $\mathcal{P}(\Omega)$. However, there are mathematical reasons showing that $\mathcal{P}(\Omega)$ is in general too big. But one can also prove that,

for a given collection $\mathcal{C}$ of subsets of Ω, there exists a smallest σ-field $\sigma(\mathcal{C})$ on Ω containing $\mathcal{C}$.

We call $\sigma(\mathcal{C})$ the *σ-field generated by $\mathcal{C}$.*

Example 1.4.6 (Generating σ-fields from collections of subsets of Ω)
Recall the σ-fields from Example 1.4.5. Prove that $\mathcal{F}_i = \sigma(\mathcal{C}_i)$, where

$$\mathcal{C}_1 = \{\emptyset\}, \quad \mathcal{C}_2 = \{A\}, \quad \mathcal{C}_3 = \mathcal{F}_3.$$

Using the definition of a σ-field, you have to check which sets *necessarily* belong to the σ-field $\sigma(\mathcal{C}_i)$.

Now consider

$$\mathcal{C}_4 = \{A, B\} \quad \text{and} \quad \mathcal{C}_5 = \{A, B, C\},$$

where $A, B, C \subset \Omega$ and determine $\sigma(\mathcal{C}_4)$ and $\sigma(\mathcal{C}_5)$. Before you start: first think about the structure of the elements in $\sigma(\mathcal{C}_4)$ and $\sigma(\mathcal{C}_5)$. Notice that you can get every element of $\sigma(\mathcal{C}_4)$ by taking all possible unions of the sets

$$\emptyset, \quad A \cap B, \quad A^c \cap B, \quad A \cap B^c, \quad A^c \cap B^c.$$

For $\sigma(\mathcal{C}_5)$ you can proceed in a similar way. □

In general, it is difficult, if not impossible, to give a constructive description of the elements in $\sigma(C)$. The σ-field $\sigma(Y)$ generated by a discrete random variable Y is an exception.

Example 1.4.7 (The σ-field generated by a discrete random variable)
Recall the set-up of Section 1.4.1. We considered a discrete random variable Y with distinct values y_i and defined the subsets $A_i = \{\omega : Y(\omega) = y_i\}$, which constitute a disjoint partition of Ω. Choose

$$C = \{A_1, A_2, \ldots\}.$$

Since $\sigma(C)$ is a σ-field it must contain all sets of the form

$$A = \bigcup_{i \in I} A_i, \tag{1.32}$$

where I is any subset of $\mathbb{N} = \{1, 2, \ldots\}$, including $I = \emptyset$ (giving $A = \emptyset$) and $I = \mathbb{N}$ (giving $A = \Omega$). Verify that the sets (1.32) constitute a σ-field $\sigma(Y)$. Since the sets (1.32) *necessarily* belong to $\sigma(C)$, the smallest σ-field containing C, we also have $\sigma(Y) = \sigma(C)$. Later we will call $\sigma(Y)$ the *σ-field generated by Y*.

Notice that $\sigma(Y)$ contains all sets of the form

$$A_{a,b} = \{Y \in (a, b]\} = \{\omega : a < Y(\omega) \le b\}, \quad -\infty < a < b < \infty.$$

Indeed, the set $I = \{i : a < y_i \le b\}$ is a subset of $\mathbb{N}$, hence

$$A_{a,b} = \bigcup_{i \in I} \{\omega : Y(\omega) = y_i\} \in \sigma(Y). \qquad \square$$

Example 1.4.8 (The Borel sets)
Take $\Omega = \mathbb{R}$ and

$$C^{(1)} = \{(a, b] : -\infty < a < b < \infty\}.$$

The σ-field $\mathcal{B}_1 = \sigma(C^{(1)})$ contains very general subsets of $\mathbb{R}$. It is called the *Borel σ-field*, its elements are the *Borel sets*. A normal human being cannot imagine the large variety of Borel sets. For example, it is a true fact, but not easy to verify, that $\mathcal{B}_1$ is a genuine subset of the power set $\mathcal{P}(\mathbb{R})$.

One can also introduce the σ-field of the *n-dimensional Borel sets* $\mathcal{B}_n = \sigma(C^{(n)})$, where $\Omega = \mathbb{R}^n$ and

$$C^{(n)} = \{(\mathbf{a}, \mathbf{b}] : -\infty < a_i < b_i < \infty, \quad i = 1, \ldots, n\}.$$

The sets $(\mathbf{a}, \mathbf{b}]$ are called *rectangles*. Any "reasonable" subset of $\mathbb{R}^n$ is a Borel set. For example, balls, spheres, smooth curves, surfaces, etc., are Borel sets, and so are the open and the closed sets. In order to show that a given subset $C \subset \mathbb{R}^n$ is a Borel set, it is necessary to obtain C by a countable number of operations $\cap, \cup,^c$ acting on the rectangles.

For example, show for $n = 1$ that every point set $\{a\}$, $a \in \mathbb{R}$, is a Borel set. Also check that the intervals (a, b), $[a, b)$ $(-\infty, a)$, (b, ∞) are Borel sets. □

Now recall Example 1.4.7, where we generated the σ-field $\sigma(Y)$ from the sets $A_i = \{\omega : Y(\omega) = y_i\}$ for a discrete random variable Y with values y_i. If Y is a random variable assuming a continuum of values, the σ-field generated from the sets $\{\omega : Y(\omega) = y\}$, $y \in \mathbb{R}$, is, from a mathematical point of view, not rich enough. For example, sets of the form $\{\omega : a < Y(\omega) \le b\}$ do not belong to such a σ-field, but it is desirable to have them in $\sigma(Y)$.

We saw in Example 1.4.7 that, for a discrete random variable Y, the sets $\{\omega : a < Y(\omega) \le b\}$ belong to $\sigma(Y)$. Therefore we will require as a minimal assumption that these sets belong to $\sigma(Y)$ when Y is any random variable. Since we are also interested in random vectors $\mathbf{Y}$, we immediately define the σ-fields $\sigma(\mathbf{Y})$ for the multivariate case:

Let $\mathbf{Y} = (Y_1, \ldots, Y_n)$ be an n-dimensional random vector, i.e. $Y_1, \ldots, Y_n$ are random variables. The σ-field $\sigma(\mathbf{Y})$ is the smallest σ-field containing all sets of the form

$$\{\mathbf{Y} \in (\mathbf{a}, \mathbf{b}]\} = \{\omega : a_i < Y_i(\omega) \le b_i, \quad i = 1, \ldots, n\},$$

$$-\infty < a_j < b_j < \infty, \quad j = 1, \ldots, n.$$

We call $\sigma(\mathbf{Y})$ the *σ-field generated by the random vector* $\mathbf{Y}$.

An element of $\sigma(\mathbf{Y})$ tells us for which $\omega \in \Omega$ the random vector $\mathbf{Y}$ assumes values in a rectangle $(\mathbf{a}, \mathbf{b}]$ or in a more general Borel set.

The σ-field $\sigma(\mathbf{Y})$ generated by $\mathbf{Y}$ contains the essential information about the structure of the random vector $\mathbf{Y}$ as a function of $\omega \in \Omega$. It contains all sets of the form $\{\omega : \mathbf{Y}(\omega) \in C\}$ for all Borel sets $C \subset \mathbb{R}^n$.

You will admit that it is difficult to imagine the σ-field $\sigma(\mathbf{Y})$, and it is even more complicated for a stochastic process.

> For a stochastic process $Y = (Y_t, t \in T, \omega \in \Omega)$, the σ-field $\sigma(Y)$ is the smallest σ-field containing all sets of the form
>
> $$\{\omega : \text{the sample path } (Y_t(\omega), t \in T) \text{ belongs to } C\}$$
>
> for all suitable sets C of functions on T. Then $\sigma(Y)$ is called *the σ-field generated by Y*.

In view of our restricted tools, we cannot make this definition more precise, but we will support our imagination about the σ-field $\sigma(Y)$ by considering an example:

Example 1.4.9 Let $B = (B_s, s \le t)$ be Brownian motion on $[0, t]$, see Section 1.3.1, and

$$\mathcal{F}_t = \sigma(B) = \sigma(B_s, s \le t).$$

This is the smallest σ-field containing the essential information about the structure of the process B on $[0, t]$. One can show that this σ-field is generated by all sets of the form

$$A_{t_1,\dots,t_n}(C) = \{\omega : (B_{t_1}(\omega), \dots, B_{t_n}(\omega)) \in C\}$$

for any n-dimensional Borel set C, any choices of $t_i \in [0, t]$, $n \ge 1$. $\qquad\square$

We learnt that the σ-field $\sigma(Y)$ is a non-trivial object. In what follows, we want to use the following rule of thumb in order to be able to imagine $\sigma(Y)$:

> For a random variable, a random vector or a stochastic process Y on Ω, the σ-field $\sigma(Y)$ generated by Y contains the essential information about the structure of Y as a function of $\omega \in \Omega$. It consists of all subsets $\{\omega : Y(\omega) \in C\}$ for suitable sets C.
>
> Because Y generates a σ-field, we also say that Y *contains information represented by $\sigma(Y)$* or Y *carries the information $\sigma(Y)$*.

We conclude with a useful remark. Let f be a function acting on Y, and consider the set

$$\{\omega : f(Y(\omega)) \in C\}$$

for suitable sets C. For "nice" functions f, this set belongs to $\sigma(Y)$, i.e.

$$\sigma(f(Y)) \subset \sigma(Y).$$

This means that a function f acting on Y does not provide new information about the structure of Y. We say that the information carried by $f(Y)$ is contained in the information $\sigma(Y)$. We give a simple example:

Example 1.4.10 As before, let B be Brownian motion and define the σ-fields

$$\mathcal{F}_t = \sigma(B_s, s \leq t), \quad t \geq 0.$$

Consider the function $f(B) = B_t$ for a fixed t. Given that we know the structure of the whole process $(B_s, s \leq t)$ on Ω, then we also know the structure of the random variable B_t, thus $\sigma(B_t) \subset \mathcal{F}_t$. The converse is clearly not true. If we know the random variable B_t we cannot reconstruct the whole process $(B_s, s \leq t)$ from it. □

1.4.3 The General Conditional Expectation

In Section 1.4.1 we defined the conditional expectation $E(X \mid Y)$ of a random variable X given the discrete random variable Y. This definition does not make explicit use of the values y_i of Y, but it depends on the subsets $A_i = \{\omega : Y(\omega) = y_i\}$ of Ω. We learnt in Example 1.4.7 that the collection of the sets A_i generates the σ-field $\sigma(Y)$. This is the starting point for the definition of the general conditional expectation $E(X \mid \mathcal{F})$ given a σ-field $\mathcal{F}$ on Ω. In applications we will always choose $\mathcal{F} = \sigma(Y)$ for appropriate random variables, random vectors or stochastic processes Y. In the previous section we tried to convince you that the essential information about the structure of Y is contained in the σ-field $\sigma(Y)$, generated by Y. In this sense, we say that Y *carries the information* $\sigma(Y)$.

Let Y, Y_1, Y_2 be random variables, random vectors or stochastic processes on Ω and $\mathcal{F}$ be a σ-field on Ω.

We say that

- *the information of Y is contained in $\mathcal{F}$ or Y does not contain more information than that contained in $\mathcal{F}$ if $\sigma(Y) \subset \mathcal{F}$, and*

- *Y_2 contains more information than Y_1 if $\sigma(Y_1) \subset \sigma(Y_2)$.*

Now we are prepared to give a rigorous definition of the conditional expectation $E(X \mid \mathcal{F})$ under an abstract σ-field $\mathcal{F}$:

A random variable Z is called the *conditional expectation of X given the σ-field* $\mathcal{F}$ (we write $Z = E(X \mid \mathcal{F})$) if

- Z does not contain more information than that contained in $\mathcal{F}$: $\sigma(Z) \subset \mathcal{F}$.

- Z satisfies the relation

$$E(X I_A) = E(Z I_A) \quad \text{for all} \quad A \in \mathcal{F}. \qquad (1.33)$$

In Appendix A6 we show the existence and uniqueness of $E(X \mid \mathcal{F})$ by measure-theoretic means.

   Property (1.33) shows that the random variables X and $E(X \mid \mathcal{F})$ are "close" to each other, not in the sense that they coincide for any ω, but averages (expectations) of X and $E(X \mid \mathcal{F})$ on suitable sets A are the same. This supports our imagination about the random variable $E(X \mid \mathcal{F})$ which we gained in Section 1.4.1:

The conditional expectation $E(X \mid \mathcal{F})$ is a coarser version of the original random variable X.

We mention that the defining property (1.33) leaves us some freedom for the construction of the random variable $Z = E(X \mid \mathcal{F})$. This means that there can exist versions Z' of $E(X \mid \mathcal{F})$ which can differ from Z on sets of probability 0. For this reason, all relations involving $E(X \mid \mathcal{F})$ should actually be interpreted in an almost sure sense, and we should also indicate this uncertainty in all relevant formulae with the label "a.s.". However, once having pointed out this fact, we will find it convenient to suppress this label everywhere.

Example 1.4.11 (Conditional expectation under discrete condition)
We want to show that the definition of $E(X \mid Y)$ in Section 1.4.1 yields a special case of the above definition of $E(X \mid \mathcal{F})$ when $\mathcal{F} = \sigma(Y)$. Recall from Example 1.4.7 that every element A of $\sigma(Y)$ is of the form

$$A = \bigcup_{i \in I} A_i = \bigcup_{i \in I} \{\omega : Y(\omega) = y_i\}, \quad I \subset \mathbb{N}. \qquad (1.34)$$

In (1.29) we defined $E(X \mid Y)$ as the random variable Z such that

$$Z(\omega) = E(X \mid A_i) \quad \text{for } \omega \in A_i.$$

We learnt on p. 62 that Z is merely a function of Y, not of X, hence $\sigma(Z) \subset \sigma(Y)$; see the discussion at the end of Section 1.4.2. Moreover, for A given by (1.34),

$$E(XI_A) = E\left(X \sum_{i \in I} I_{A_i}\right) = \sum_{i \in I} E(XI_{A_i}).$$

On the other hand, we see that ZI_A is a discrete random variable with expectation

$$E(ZI_A) = \sum_{i \in I} E(X \mid A_i) P(A_i) = \sum_{i \in I} E(XI_{A_i}).$$

Thus Z satisfies the defining relation (1.33) and it does not contain more information than Y (it is a function of Y). Therefore it is indeed the conditional expectation of X given the σ-field $\sigma(Y)$. □

In the case of a discrete random variable Y we have just learnt that $E(X \mid Y)$ and $E(X \mid \sigma(Y))$ represent the same random variable. This suggests the following definition:

Let Y be a random variable, a random vector or a stochastic process on Ω and $\sigma(Y)$ the σ-field generated by Y.

The *conditional expectation of a random variable X given Y* is defined by

$$E(X \mid Y) = E(X \mid \sigma(Y)).$$

Example 1.4.12 (The classical conditional probability and conditional expectation)

The classical conditional probability and conditional expectation are special cases of the general notion of conditional expectation defined on p. 68. Indeed, let B be such that $P(B) > 0$, $P(B^c) > 0$ and define $\mathcal{F}_B = \sigma(\{B\})$. We know from Example 1.4.6 that $\mathcal{F}_B = \{\emptyset, \Omega, B, B^c\}$. An appeal to Example 1.4.11 yields that

$$E(X \mid \mathcal{F}_B)(\omega) = E(X \mid B) \quad \text{for } \omega \in B.$$

This is the classical notion of conditional expectation. If we specify $X = I_A$ for some event A, we obtain for $\omega \in B$,

$$E(I_A \mid \mathcal{F}_B)(\omega) = E(I_A \mid B) = \frac{P(A \cap B)}{P(B)}.$$

The right-hand side is the classical conditional probability of A given B. □

1.4.4 Rules for the Calculation of Conditional Expectations

The defining property (1.33) of $E(X \mid \mathcal{F})$ is not a constructive one. Therefore it is in general difficult, if not impossible, to calculate $E(X \mid \mathcal{F})$. The case $\mathcal{F} = \sigma(Y)$ for a discrete random variable Y, which was discussed in Example 1.4.11, is an exception. For this reason, it is important to be able to work with conditional expectations *without knowing their particular forms*. This means we have to know a few rules in order to deal with conditional expectations in standard situations.

In what follows, we collect the most common rules. In some cases we will give reasons for these rules, in the other situations we have to rely on intuitive arguments. We start with an existence and uniqueness result.

If $E|X| < \infty$, the conditional expectation $E(X \mid \mathcal{F})$ *exists* and is *unique* in the sense as discussed in Appendix A6.

In Section 1.4.1 we encountered rules for the calculation of conditional expectations under a discrete condition. They remain valid in the general case.

Rule 1

The conditional expectation is *linear*: for random variables X_1, X_2 and constants c_1, c_2,

$$E([c_1 X_1 + c_2 X_2] \mid \mathcal{F}) = c_1 E(X_1 \mid \mathcal{F}) + c_2 E(X_2 \mid \mathcal{F}). \qquad (1.35)$$

This follows by an application of the defining property (1.33) to the right- and left-hand sides of (1.35) (try to prove it!).

Now take $A = \Omega$ in property (1.33). You immediately obtain:

Rule 2

The expectations of X and $E(X \mid \mathcal{F})$ are the same:

$$EX = E[E(X \mid \mathcal{F})].$$

Also the third claim (see (1.30)) carries over from the discrete case:

> **Rule 3**
>
> If X and the σ-field $\mathcal{F}$ are independent, then $E(X \mid \mathcal{F}) = EX$.
>
> In particular, if X and Y are independent, then $E(X \mid Y) = EX$.

This statement needs some discussion: what does independence between X and $\mathcal{F}$ mean? It means that we do not gain any information about X, if we know $\mathcal{F}$, and vice versa. More formally: the random variables X and I_A are independent for all $A \in \mathcal{F}$. Now, by independence,

$$E(XI_A) = EX \, EI_A = EX \, P(A) = E[(EX) I_A] , \quad A \in \mathcal{F} .$$

A comparison with (1.33) yields that the constant random variable $Z = EX$ is the conditional expectation $E(X \mid \mathcal{F})$. This proves Rule 3.

> **Rule 4**
>
> If the σ-field $\sigma(X)$, generated by the random variable X, is contained in $\mathcal{F}$, then
>
> $$E(X \mid \mathcal{F}) = X .$$
>
> In particular, if X is a function of Y, $\sigma(X) \subset \sigma(Y)$, thus $E(X \mid Y) = X$.

This means that the information contained in $\mathcal{F}$ provides us with the whole information about the random variable X. If we know everything about the structure of X, we can deal with it *as if it was non-random* and write the value $X(\omega)$ in front of the conditional expectation $E(1 \mid \mathcal{F}) = 1$:

$$E(X \mid \mathcal{F})(\omega) = E(X(\omega) \mid \mathcal{F}) = X(\omega) \, E(1 \mid \mathcal{F}) = X(\omega) .$$

This rule can be extended to more general situations:

> **Rule 5**
>
> If the σ-field $\sigma(X)$, generated by the random variable X, is contained in $\mathcal{F}$, then for any random variable G,
>
> $$E(XG \mid \mathcal{F}) = X \, E(G \mid \mathcal{F}) .$$
>
> In particular, if X is a function of Y, $\sigma(X) \subset \sigma(Y)$, thus $E(XG \mid Y) = XE(G \mid Y)$.

Indeed, given $\mathcal{F}$, we can deal with X as if it was a constant, hence we can pull $X(\omega)$ out of the updated expectation and write it in front of $E(G \mid \mathcal{F})$.

Rule 6

If $\mathcal{F}$ and $\mathcal{F}'$ are two σ-fields with $\mathcal{F} \subset \mathcal{F}'$, then

$$E(X \mid \mathcal{F}) \;=\; E(E(X \mid \mathcal{F}') \mid \mathcal{F}), \qquad (1.36)$$

$$E(X \mid \mathcal{F}) \;=\; E(E(X \mid \mathcal{F}) \mid \mathcal{F}'). \qquad (1.37)$$

For obvious reasons, Rule 6 is sometimes called *the tower property of conditional expectations*.

Rule (1.37) can be justified by Rule 4: since $\mathcal{F} \subset \mathcal{F}'$, $E(X \mid \mathcal{F})$ does not contain more information than $\mathcal{F}'$, i.e. given $\mathcal{F}'$, we can deal with $E(X \mid \mathcal{F})$ like a constant:

$$E(E(X \mid \mathcal{F}) \mid \mathcal{F}') = E(X \mid \mathcal{F}) \, E(1 \mid \mathcal{F}') = E(X \mid \mathcal{F}).$$

Rule (1.36) can formally be derived from the defining property (1.33) which says that for $A \in \mathcal{F}$ and $Z = E(X \mid \mathcal{F})$,

$$E(X I_A) = E(Z I_A). \qquad (1.38)$$

On the other hand, by Rule 5 and since $A \in \mathcal{F} \subset \mathcal{F}'$,

$$E(E(X \mid \mathcal{F}') \mid \mathcal{F}) \, I_A = E(E(X \mid \mathcal{F}')I_A \mid \mathcal{F}) = E(E(X I_A \mid \mathcal{F}') \mid \mathcal{F}).$$

Now take expectations and apply Rule 2 to the right-hand side:

$$E[E(E(X \mid \mathcal{F}') \mid \mathcal{F}) \, I_A] = E[X I_A].$$

Thus $Z' = E(E(X \mid \mathcal{F}') \mid \mathcal{F})$ also satisfies (1.38), but since $E(X \mid \mathcal{F})$ is unique, we must have $Z = Z'$, which proves (1.36).

We finish with a generalization of Rule 3.

Rule 7

If X is independent of $\mathcal{F}$, and the information carried by the random variable, the random vector or the stochastic process G is contained in $\mathcal{F}$, then for any function $h(x, y)$,

$$E(h(X, G) \mid \mathcal{F}) = E(E_X[h(X, G)] \mid \mathcal{F}),$$

where $E_X[h(X, G)]$ means that we fix G and take the expectation with respect to X.

We illustrate Rule 7 by an example.

Example 1.4.13 Let X and Y be independent random variables. Then Rules 7 and 5 give

$$E(XY \mid Y) = E(E_X(XY) \mid Y) = E(YEX \mid Y) = YEX \,,$$

$$E(X + Y \mid Y) = E(E_X(X + Y) \mid Y) = E(EX + Y \mid Y) = EX + Y \,. \qquad \square$$

In the following two examples we want to exercise the rules for calculating conditional expectations.

Example 1.4.14 (Brownian motion)
Recall the definition of Brownian motion $B = (B_t, t \geq 0)$ from p. 33. We associate with B an increasing stream of information about the structure of the process which is represented by the σ-fields $\mathcal{F}_s = \sigma(B_x, x \leq s)$. We want to calculate

$$E(B_t \mid \mathcal{F}_s) = E(B_t \mid B_x, x \leq s) \quad \text{for} \quad s \geq 0 \,.$$

Clearly, if $s \geq t$, $\mathcal{F}_s \supset \mathcal{F}_t$, and so Rule 4 gives

$$E(B_t \mid \mathcal{F}_s) = B_t \,.$$

Now assume $s < t$. Then, by linearity of the conditional expectation (Rule 1),

$$
\begin{aligned}
E(B_t \mid \mathcal{F}_s) &= E[(B_t - B_s) + B_s \mid \mathcal{F}_s] \\
&= E(B_t - B_s \mid \mathcal{F}_s) + E(B_s \mid \mathcal{F}_s) \,.
\end{aligned}
$$

Since $B_t - B_s$ is independent of $\mathcal{F}_s$, Rule 3 applies:

$$E(B_t - B_s \mid \mathcal{F}_s) = E(B_t - B_s) = 0 \,.$$

Moreover, $\sigma(B_s) \subset \sigma(B_x, x \leq s) = \mathcal{F}_s$, hence $E(B_s \mid \mathcal{F}_s) = B_s$. Finally,

$$E(B_t \mid \mathcal{F}_s) = B_{\min(s,t)} \,. \qquad \square$$

Example 1.4.15 (Squared Brownian motion)
As in Example 1.4.14, B denotes Brownian motion and $\mathcal{F}_s = \sigma(B_x, x \leq s)$. We consider the stochastic process $X_t = B_t^2 - t$, $t \geq 0$. The same arguments as in Example 1.4.14 yield

$$E(X_t \mid \mathcal{F}_s) = X_t \quad \text{for} \quad s \geq t \,.$$

For $s < t$, we write

$$
\begin{aligned}
B_t^2 - t &= [(B_t - B_s) + B_s]^2 - t \\
&= (B_t - B_s)^2 + B_s^2 + 2B_s(B_t - B_s) - t.
\end{aligned}
$$

Taking conditional expectations, we obtain

$$
E(X_t \mid \mathcal{F}_s) = E[(B_t - B_s)^2 \mid \mathcal{F}_s] + E(B_s^2 \mid \mathcal{F}_s) + 2E[B_s(B_t - B_s) \mid \mathcal{F}_s] - t.
$$

Notice that $B_t - B_s$ and $(B_t - B_s)^2$ are independent of $\mathcal{F}_s$, and that $\sigma(B_s^2) \subset \sigma(B_s) \subset \mathcal{F}_s$. Applying Rules 3–5, we obtain

$$
\begin{aligned}
E(X_t \mid \mathcal{F}_s) &= E(B_t - B_s)^2 + B_s^2 + 2B_s\, E(B_t - B_s) - t \\
&= (t - s) + B_s^2 + 0 - t = X_s.
\end{aligned}
$$

Finally,

$$
E(X_t \mid \mathcal{F}_s) = X_{\min(s,t)}.
$$

Later we will take this relation as the defining property for a martingale; see (1.41). $\qquad\square$

1.4.5 The Projection Property of Conditional Expectations

In what follows, $\mathcal{F}$ is a σ-field and $L^2(\mathcal{F})$ is the collection of random variables Z on Ω, satisfying the following conditions:

- Z has a finite second moment: $EZ^2 < \infty$,

- The information carried by Z is contained in $\mathcal{F}$: $\sigma(Z) \subset \mathcal{F}$. If $\mathcal{F} = \sigma(Y)$ this means that Z is a function of Y.

The random variable $E(X \mid \mathcal{F})$ can be understood as an updated version of the expectation of X, given the information $\mathcal{F}$. The conditional expectation has a certain optimality property in the class $L^2(\mathcal{F})$:

The Projection Property

Let X be a random variable with $EX^2 < \infty$. The conditional expectation $E(X \mid \mathcal{F})$ is that random variable in $L^2(\mathcal{F})$ which is closest to X in the mean square sense. This means that

$$
E[X - E(X \mid \mathcal{F})]^2 = \min_{Z \in L^2(\mathcal{F})} E(X - Z)^2. \tag{1.39}
$$

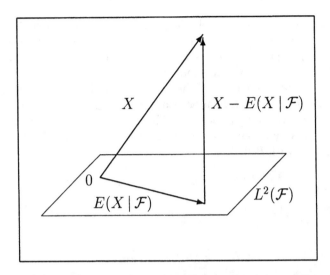

Figure 1.4.16 *An illustration of the projection property of the conditional expectation $E(X\,|\,\mathcal{F})$. We mention that $< Z, Y > = E(ZY)$ for Z, Y with $EZ^2 < \infty$ and $EY^2 < \infty$ defines an inner product and $\|Z - Y\| = \sqrt{< Z - Y, Z - Y >}$ a distance between Z and Y. As in Euclidean space, we say that Z and Y are orthogonal if $< Z, Y > = 0$. In this sense, $E(X\,|\,\mathcal{F})$ is the orthogonal projection of X on $L^2(\mathcal{F})$: $< X - E(X\,|\,\mathcal{F}), Z > = 0$ for all $Z \in L^2(\mathcal{F})$, and $< X - Z, X - Z >$ is minimal for $Z = E(X\,|\,\mathcal{F})$.*

In this sense, $E(X\,|\,\mathcal{F})$ is the *projection of the random variable X on the space* $L^2(\mathcal{F})$ of the random variables Z carrying part of the information $\mathcal{F}$; also see the comments to Figure 1.4.16.

If $\mathcal{F} = \sigma(Y)$, $E(X\,|\,Y)$ is that function of Y which has a finite second moment and which is closest to X in the mean square sense.

In what follows, we sometimes refer to $E(X\,|\,\mathcal{F})$ as the *best prediction of X given $\mathcal{F}$*. This means that relation (1.39) holds. To give some meaning to the word "prediction", we reconsider Examples 1.4.14 and 1.4.15. We proved that for $s \le t$,

$$E(B_t\,|\,B_x, x \le s) = B_s \quad \text{and} \quad E(B_t^2 - t\,|\,B_x, x \le s) = B_s^2 - s\,.$$

Thus the best predictions of the future values B_t and $B_t^2 - t$, given the information about Brownian motion until the present time s, are the present values B_s and $B_s^2 - s$, respectively. This property characterizes the whole class of martingales with a finite second moment: the best prediction of the future values of the stochastic process is the present value; see Section 1.5.1.

In what follows, we indicate the steps for the proof of the projection property (1.39). We only make use of the rules for calculating conditional expectations. First notice that the random variable $Z' = E(X \mid \mathcal{F})$ belongs to $L^2(\mathcal{F})$: it carries information only about $\mathcal{F}$ and has a finite second moment. The latter property follows by an application of *Jensen's inequality* (see (A.2) on p. 188) in combination with Rule 2:

$$E[(E(X \mid \mathcal{F}))^2] \le E[E(X^2 \mid \mathcal{F})] = EX^2 .$$

Now we are going to check the projection property (1.39). Let Z be any random variable in $L^2(\mathcal{F})$. Then

$$
\begin{aligned}
E(X - Z)^2 &= E((X - Z') + (Z' - Z))^2 \\
&= E(X - Z')^2 + E(Z' - Z)^2 + 2\, E[(X - Z')(Z' - Z)] .
\end{aligned}
$$

We treat the terms on the right-hand side separately. First consider $E[(X - Z')(Z' - Z)]$. Since both, Z and Z', belong to $L^2(\mathcal{F})$, so does $Z - Z'$. In particular, by Rule 5,

$$E[(X - Z')(Z' - Z) \mid \mathcal{F}] = (Z - Z')\, E(X - Z' \mid \mathcal{F}) .$$

But by Rules 1 and 4,

$$E(X - Z' \mid \mathcal{F}) = E(X \mid \mathcal{F}) - E(Z' \mid \mathcal{F}) = Z' - Z' = 0 .$$

Thus we proved

$$E(X - Z)^2 = E(X - Z')^2 + E(Z' - Z)^2 .$$

Hence,

$$E(X - Z)^2 \ge E(X - Z')^2 \quad \text{for all random variables } Z \text{ in } L^2(\mathcal{F}). \qquad (1.40)$$

We also see that equality in (1.40) is achieved if $Z = Z'$, so $Z' = E(X \mid \mathcal{F})$ really represents that element of $L^2(\mathcal{F})$, for which the minimum of $E(X - Z)^2$ is attained. This proves (1.39).

Notes and Comments

The notion of conditional expectation is one of the most difficult ones in probability theory, but it is also one of the most powerful tools. Its definition as a random variable given a σ-field goes back to Kolmogorov. For its complete theoretical understanding, measure-theoretic probability theory is unavoidable.

The conditional expectation is treated in every advanced textbook on probability theory, see for example Billingsley (1995) or Williams (1991).

1.5 Martingales

1.5.1 Defining Properties

The notion of a martingale is crucial for the understanding of the Itô stochastic integral. Indefinite Itô stochastic integrals are constructed in such a way that they constitute martingales. The idea underlying a martingale is a fair game where the net winnings are evaluated via conditional expectations. Fortunately, we now have this powerful instrument in our tool box; see Section 1.4.

Assume that $(\mathcal{F}_t, t \geq 0)$ is a collection of σ-fields on the same space Ω and that all $\mathcal{F}_t$s are subsets of a larger σ-field $\mathcal{F}$ on Ω, say.

The collection $(\mathcal{F}_t, t \geq 0)$ of σ-fields on Ω is called a *filtration* if

$$\mathcal{F}_s \subset \mathcal{F}_t \quad \text{for all } 0 \leq s \leq t.$$

Thus a filtration is an increasing stream of information.

If $(\mathcal{F}_n, n = 0, 1, \ldots)$ is a sequence of σ-fields on Ω and $\mathcal{F}_n \subset \mathcal{F}_{n+1}$ for all n, we call $(\mathcal{F}_n)$ a *filtration* as well.

For our applications, a filtration is usually linked up with a stochastic process:

The stochastic process $Y = (Y_t, t \geq 0)$ is said to be *adapted to the filtration* $(\mathcal{F}_t, t \geq 0)$ if

$$\sigma(Y_t) \subset \mathcal{F}_t \quad \text{for all } t \geq 0.$$

The stochastic process Y is always adapted to the *natural filtration generated by* Y:

$$\mathcal{F}_t = \sigma(Y_s, s \leq t).$$

Thus adaptedness of a stochastic process Y means that the Y_ts do not carry more information than $\mathcal{F}_t$.

If $Y = (Y_n, n = 0, 1, \ldots)$ is a discrete-time process we define adaptedness in an analogous way: for a filtration $(\mathcal{F}_n, n = 0, 1, \ldots)$ we require that $\sigma(Y_n) \subset \mathcal{F}_n$.

Example 1.5.1 (Examples of adapted processes)
Let $(B_t, t \geq 0)$ be Brownian motion and $(\mathcal{F}_t, t \geq 0)$ be the corresponding natural filtration. Stochastic processes of the form

$$X_t = f(t, B_t), \quad t \geq 0,$$

where f is a function of two variables, are adapted to $(\mathcal{F}_t, t \geq 0)$. This includes the processes

$$X_t^{(1)} = B_t, \quad X_t^{(2)} = B_t^2, \quad X_t^{(3)} = B_t^2 - t, \quad X_t^{(4)} = B_t^3, \quad X_t^{(5)} = B_t^4.$$

But also processes, which may depend on the whole past of Brownian motion, can be adapted. For example,

$$X_t^{(6)} = \max_{0 \leq s \leq t} B_s \quad \text{or} \quad X_t^{(7)} = \min_{0 \leq s \leq t} B_s^2.$$

If the stochastic process Y is adapted to the natural Brownian filtration $(\mathcal{F}_t, t \geq 0)$, we will say that Y is *adapted to Brownian motion*. This means that Y_t is a function of B_s, $s \leq t$.

The following processes are *not adapted* to Brownian motion:

$$X_t^{(8)} = B_{t+1}, \quad X_t^{(9)} = B_T - B_t, \quad X_t^{(10)} = B_t + B_T,$$

where $T > 0$ is a fixed number. Indeed, these processes require the knowledge of Brownian motion at future instants of time. For example, consider $X_t^{(9)}$. For its definition you have to know B_T at times $t < T$. $\square$

Clearly, one can enlarge the natural filtration and obtain another filtration to which the stochastic process is adapted.

Example 1.5.2 (Enlarging a filtration)
Consider Brownian motion $B = (B_t, t \geq 0)$ and the corresponding natural filtration $\mathcal{F}_t = \sigma(B_s, s \leq t)$, $t \geq 0$. The stochastic process $X_t = B_t^2$ generates the natural filtration

$$\mathcal{F}_t' = \sigma(B_s^2, s \leq t), \quad t \geq 0,$$

which is smaller than $(\mathcal{F}_t)$. Indeed, for every t, $\mathcal{F}_t' \subset \mathcal{F}_t$, since we can only reconstruct the whole information about $|B_t|$ from B_t^2, but not about B_t: we can say nothing about the sign of B_t. Then $(\mathcal{F}_t)$ is also a filtration for (B_t^2).

Thus we can work with different filtrations for the same process. To illustrate this aspect from an applied point of view, we consider an example from stochastic finance. In this field, it is believed that share prices, exchange rates, interest rates, etc., can be modelled by solutions of stochastic differential equations which are driven by Brownian motion; see Chapter 4. These solutions are then functions of Brownian motion. This process models the fluctuations of the financial market (independent movements up and down on disjoint time intervals). These fluctuations actually represent the information about the market. This relevant knowledge is contained in the natural filtration. It does not take information from outside into account. However, in finance there are always people who know more than the others. For example, they might know that an essential political decision will be taken in the very near future which will completely change the financial landscape. This enables the informed persons to act with more competence than the others. Thus they have their own filtrations which can be bigger than the natural filtration.   □

Now consider a stochastic process $X = (X_t, t \geq 0)$ on Ω and suppose you have the information $\mathcal{F}_s$ at the present time s.

How does this information influence our knowledge about the behavior of the process X in the future?

If $\mathcal{F}_s$ and X are dependent, we can expect that our information reduces the uncertainty about the values of X_t at a future instant of time t. If we know that certain events happened in the past, we may include this knowledge in our calculations. Thus X_t can be better predicted with the information $\mathcal{F}_s$ than without it. A mathematical tool to describe this gain of information is the conditional expectation of X_t given $\mathcal{F}_s$:

$$E(X_t \,|\, \mathcal{F}_s) \quad \text{for} \quad 0 \leq s < t.$$

We learnt in Section 1.4.5 that $E(X_t \,|\, \mathcal{F}_s)$ is the best prediction of X_t given the information $\mathcal{F}_s$. Also recall from Examples 1.4.14 and 1.4.15 that the stochastic processes $X_t = B_t$ and $X_t = B_t^2 - t$ (B is Brownian motion) satisfy the condition $E(X_t \,|\, \mathcal{F}_s) = X_s$ for $s < t$. For these processes X, the best prediction of the future value X_t given $\mathcal{F}_s$ is the present value X_s. Clearly, this may change if we consider another filtration, and therefore we must always say which filtration we consider.

The stochastic process $X = (X_t, t \geq 0)$ is called a *continuous-time mar-tingale with respect to the filtration* $(\mathcal{F}_t, t \geq 0)$, we write $(X, (\mathcal{F}_t))$, if

- $E|X_t| < \infty$ for all $t \geq 0$.

- X is adapted to $(\mathcal{F}_t)$; see p. 77 for the definition.

-
$$E(X_t \,|\, \mathcal{F}_s) = X_s \quad \text{for all } 0 \leq s < t, \tag{1.41}$$

  i.e. X_s is the best prediction of X_t given $\mathcal{F}_s$.

It is also possible to define a discrete-time martingale $X = (X_n, n = 0, 1, \ldots)$. In this case, we adapt the defining property (1.41) as follows:

$$E(X_{n+k} \,|\, \mathcal{F}_n) = X_n, \quad k \geq 0. \tag{1.42}$$

We show that it suffices to require (1.42) for $k = 1$. Indeed, recalling Rule 6 from p. 72,

$$
\begin{aligned}
E(X_{n+1} \,|\, \mathcal{F}_n) &= E[E(X_{n+2} \,|\, \mathcal{F}_{n+1}) \,|\, \mathcal{F}_n] = E(X_{n+2} \,|\, \mathcal{F}_n) \\
&= E[E(X_{n+3} \,|\, \mathcal{F}_{n+2}) \,|\, \mathcal{F}_n] = E(X_{n+3} \,|\, \mathcal{F}_n) \\
&= \quad\quad\quad \cdots \quad\quad\quad = E(X_{n+k} \,|\, \mathcal{F}_n).
\end{aligned}
$$

Now we define a martingale in the discrete-time case.

The stochastic process $X = (X_n, n = 0, 1, \ldots)$ is called a *discrete-time martingale with respect to the filtration* $(\mathcal{F}_n, n = 0, 1, \ldots)$, we write $(X, (\mathcal{F}_n))$, if

- $E|X_n| < \infty$ for all $n = 0, 1, \ldots$.

- X is adapted to $(\mathcal{F}_n)$.

-
$$E(X_{n+1} \,|\, \mathcal{F}_n) = X_n \quad \text{for all } n = 0, 1, \ldots, \tag{1.43}$$

  i.e. X_n is the best prediction of X_{n+1} given $\mathcal{F}_n$.

It is not difficult to see that the defining property (1.43) can be rewritten in the form

$$E(Y_{n+1} \,|\, \mathcal{F}_n) = 0, \quad \text{where} \quad Y_{n+1} = X_{n+1} - X_n, \quad n = 0, 1, \ldots. \tag{1.44}$$

The sequence (Y_n) is then called a *martingale difference sequence with respect to the filtration* $(\mathcal{F}_n)$.

In what follows, we often say that "$(X_t, t \geq 0)$, respectively $(X_n, n = 0, 1, \ldots)$, is a martingale" without pointing out which filtration we use. This will be clear from the context.

> A martingale has the remarkable property that its expectation function is constant.

Indeed, using the defining property $E(X_t \mid \mathcal{F}_s) = X_s$ for $s < t$ and Rule 2 on p. 70, we obtain

$$EX_s = E[E(X_t \mid \mathcal{F}_s)] = EX_t \quad \text{for all } s \text{ and } t.$$

This provides an easy way of proving that a stochastic process is not a martingale. For example, if B is Brownian motion, $EB_t^2 = t$ for all t. Hence (B_t^2) cannot be a martingale. However, we cannot use this means to prove that a stochastic process is a martingale, since a process that has a constant expectation function need not be a martingale as the following example shows: we have $EB_t^3 = 0$ for all t, but (B_t^3) is also not a martingale; see Example 1.5.5.

1.5.2 Examples

In this section we collect some simple examples of stochastic processes which have the martingale property.

Example 1.5.3 (Partial sums of independent random variables constitute a martingale)

Let (Z_n) be a sequence of independent random variables with finite expectations and $Z_0 = 0$. Consider the partial sums

$$R_n = Z_0 + \cdots + Z_n, \quad n \geq 0,$$

and the corresponding natural filtration $\mathcal{F}_n = \sigma(R_0, \ldots, R_n)$ for $n \geq 0$. Notice that

$$\mathcal{F}_n = \sigma(Z_0, \ldots, Z_n), \quad n \geq 0.$$

Indeed, the random vectors $(Z_0, \ldots, Z_n)$ and $(R_0, \ldots, R_n)$ contain the same information since $R_i = Z_1 + \cdots + Z_i$ and $Z_i = R_i - R_{i-1}$ for $i = 1, \ldots, n$.

An application of Rules 1, 3 and 4 in Section 1.4.4 yields

$$E(R_{n+1} \mid \mathcal{F}_n) = E(R_n \mid \mathcal{F}_n) + E(Z_{n+1} \mid \mathcal{F}_n) = R_n + EZ_{n+1}.$$

Hence, if $EZ_n = 0$ for all n, then $(R_n, n = 0, 1, \ldots)$ is a martingale with respect to $(\mathcal{F}_n, n = 0, 1, \ldots)$. $\qquad \square$

Example 1.5.4 (Collecting information about a random variable)
Let Z be a random variable on Ω with $E|Z| < \infty$ and $(\mathcal{F}_t, t \geq 0)$ be a filtration on Ω. Define the stochastic process X as follows:

$$X_t = E(Z \,|\, \mathcal{F}_t), \quad t \geq 0.$$

Since $\mathcal{F}_t$ increases when time goes by, X_t gives us more and more information about the random variable Z. In particular, if $\sigma(Z) \subset \mathcal{F}_t$ for some t, then $X_t = Z$. We show that X is a martingale.

An appeal to Jensen's inequality (A.2) on p. 188 and to Rule 2 on p. 70 yields

$$E|X_t| = E|E(Z \,|\, \mathcal{F}_t)| \leq E[E(|Z| \,|\, \mathcal{F}_t)] = E|Z| < \infty.$$

Moreover, X_t is obtained by conditioning on the information $\mathcal{F}_t$. Hence it does not contain more information than $\mathcal{F}_t$, so $\sigma(X_t) \subset \mathcal{F}_t$. It remains to check (1.41). Let $s < t$. Then an application of Rule 6 on p. 72 yields

$$E(X_t \,|\, \mathcal{F}_s) = E[E(Z \,|\, \mathcal{F}_t) \,|\, \mathcal{F}_s] = E(Z \,|\, \mathcal{F}_s) = X_s.$$

Thus X obeys the defining properties of a continuous-time martingale; see p. 80.   □

Example 1.5.5 (Brownian motion is a martingale)
Let $B = (B_t, t \geq 0)$ be Brownian motion. We conclude from Examples 1.4.14 and 1.4.15 that both, $(B_t, t \geq 0)$ and $(B_t^2 - t, t \geq 0)$, are martingales with respect to the natural filtration $\mathcal{F}_t = \sigma(B_s, s \leq t)$.

In the same way, you can show (do it!) that $((B_t^3 - 3tB_t), (\mathcal{F}_t))$ is a martingale.

Try to find a stochastic process (A_t) such that $((B_t^4 + A_t), (\mathcal{F}_t))$ is a martingale. Hint: first calculate $E[((B_t - B_s) + B_s)^4 \,|\, \mathcal{F}_s]$ for $s < t$.   □

The following example of a martingale transform is a first step toward the definition of the Itô stochastic integral. Indeed, such a transform can be considered as a discrete analogue of a stochastic integral.

Example 1.5.6 (Martingale transform)
Let $Y = (Y_n, n = 0, 1, \ldots)$ be a martingale difference sequence with respect to the filtration $(\mathcal{F}_n, n = 0, 1, \ldots)$; see (1.44) for the definition. Consider a stochastic process $C = (C_n, n = 1, 2, \ldots)$ and assume that, for every n, the information carried by C_n is contained in $\mathcal{F}_{n-1}$, i.e.

$$\sigma(C_n) \subset \mathcal{F}_{n-1}. \tag{1.45}$$

This means that, given $\mathcal{F}_{n-1}$, we completely know C_n at time $n - 1$.

If the sequence $(C_n, n = 1, 2, \ldots)$ satisfies (1.45), we call it *previsible* or *predictable* with respect to $(\mathcal{F}_n)$.

Now define the stochastic process

$$X_0 = 0, \quad X_n = \sum_{i=1}^{n} C_i Y_i, \quad n \geq 1. \qquad (1.46)$$

For obvious reasons, the process X is denoted by $C \bullet Y$. It is called the *martingale transform of Y by C.*

The martingale transform $C \bullet Y$ is a martingale if $EC_n^2 < \infty$ and $EY_n^2 < \infty$ for all n. We check the three defining properties on p. 80:

$$E|X_n| \leq \sum_{i=1}^{n} E|C_i Y_i| \leq \sum_{i=1}^{n} [EC_i^2 \, EY_i^2]^{1/2} < \infty,$$

where we made use of the *Cauchy–Schwarz inequality* $E|C_i Y_i| \leq [EC_i^2 \, EY_i^2]^{1/2}$ (see p. 188). Clearly, X_n is adapted to $\mathcal{F}_n$ since $Y_1, \ldots, Y_n$ do not carry more information than $\mathcal{F}_n$, and $C_1, \ldots, C_n$ is predictable. Hence $\sigma(X_n) \subset \mathcal{F}_n$. Moreover, applying Rule 5 on p. 71 and recalling that (C_n) is predictable, we obtain

$$E(X_n - X_{n-1} \,|\, \mathcal{F}_{n-1}) = E(C_n Y_n \,|\, \mathcal{F}_{n-1}) = C_n \, E(Y_n \,|\, \mathcal{F}_{n-1}) = 0.$$

In the last step we used the defining property (1.44) of the martingale difference sequence (Y_n). Hence $(X_n - X_{n-1})$ is a martingale difference sequence, and (X_n) is a martingale with respect to $(\mathcal{F}_n)$. $\qquad \square$

Example 1.5.7 (A Brownian martingale transform)
Consider Brownian motion $B = (B_s, s \leq t)$ and a partition

$$0 = t_0 < t_1 < \cdots < t_{n-1} < t_n = t.$$

Using the independent increment property of B, it is not difficult to see (check it!) that the sequence

$$\Delta B: \quad \Delta_0 B = 0, \quad \Delta_i B = B_{t_i} - B_{t_{i-1}}, \quad i = 1, \ldots, n,$$

forms a martingale difference sequence ΔB with respect to the filtration given by

$$\mathcal{F}_0 = \{\emptyset, \Omega\}, \quad \mathcal{F}_i = \sigma(B_{t_j}, 1 \leq j \leq i), \quad i = 1, \ldots, n.$$

Now consider the transforming sequence

$$\widetilde{B} = (B_{t_{i-1}}, \ i = 1, \ldots, n).$$

It is predictable with respect to $(\mathcal{F}_n)$ (why?). The martingale transform $\widetilde{B} \bullet \Delta B$ is then a martingale:

$$(\widetilde{B} \bullet \Delta B)_k = \sum_{i=1}^{k} \widetilde{B}_i \, \Delta_i B = \sum_{i=1}^{k} B_{t_{i-1}} \left(B_{t_i} - B_{t_{i-1}} \right), \quad k = 1, \ldots, n.$$

The sums on the right-hand side have the typical form of Riemann–Stieltjes sums which would be used for the definition of the Riemann–Stieltjes integral $\int_0^t B_s \, dB_s$; see Section 2.1.2. However, this integral does not exist in the Riemann–Stieltjes sense since the sample paths of Brownian motion are too irregular. We will see in Section 2.2.1 that $\widetilde{B} \bullet \Delta B$ is a discrete-time analogue of the Itô stochastic integral $\int_0^t B_s dB_s$. $\hspace{1cm} \square$

1.5.3   The Interpretation of a Martingale as a Fair Game

At the beginning of this section we mentioned that martingales are considered as models for *fair games*. This interpretation comes from the defining properties of a martingale; see p. 80.

Suppose you play a game in continuous time; i.e. at every instant of time t you have a random variable X_t as the value of the game. Suppose that (X_t) is adapted to the filtration $(\mathcal{F}_t)$. Think of $X_t - X_s$ as the *net winnings of the game per unit stake in the time frame* $(s, t]$. Then the best prediction (in the sense of Section 1.4.5) of the net winnings given the information $\mathcal{F}_s$ at time $s < t$ has value

$$E(X_t - X_s \mid \mathcal{F}_s) = E(X_t \mid \mathcal{F}_s) - X_s.$$

If $(X, (\mathcal{F}_t))$ is a martingale, the right-hand side vanishes. This means:

> The best prediction of the future net winnings per unit stake in the interval $(s, t]$ is zero.

This is exactly what you would expect to be a fair game.

People in finance do certainly gamble (although they would not admit this in public), and they even believe that they play fair. This is why they have become attracted by the world of martingales and stochastic integrals.

We know that the martingale transform $C \bullet Y$ of a martingale difference sequence Y is a martingale; see Example 1.5.6. It has an interesting interpretation in the context of fair games: think of Y_n as your net winnings per unit stake at the nth game which are adapted to a filtration $(\mathcal{F}_n)$. Your stakes C_n constitute a predictable sequence with respect to $(\mathcal{F}_n)$, i.e. at the nth game your stake C_n does not contain more information than $\mathcal{F}_{n-1}$ does. At time $n-1$ this is the best information we have about the game. The martingale transform $C \bullet Y$ gives you the net winnings per game. In particular, $(C \bullet Y)_n = \sum_{i=1}^n C_i Y_i$ are the net winnings up to time n, and $C_n Y_n$ are the net winnings per stake C_n at the nth game. It is fair since the best prediction of the net winnings $C_n Y_n$ of the nth game, just before the nth game starts, is zero: $E(C_n Y_n \mid \mathcal{F}_{n-1}) = 0$.

In Chapter 2 we will learn about the continuous-time analogue to martingale transforms: the Itô stochastic integral.

Notes and Comments

Martingales constitute an important class of stochastic processes. The theory of martingales is considered in all modern textbooks on stochastic processes. See Karatzas and Shreve (1988) and Revuz and Yor (1991) for the advanced theory of continuous-time martingales. The book by Williams (1991) contains an introduction to conditional expectations and discrete-time martingales.

2

The Stochastic Integral

In this chapter we introduce the important notion of Itô stochastic integral. We know from Section 1.3.1 that the sample paths of Brownian motion are nowhere differentiable and have unbounded variation. This has major consequences for the definition of a stochastic integral with respect to Brownian sample paths.

We discuss the notion of *ordinary integral* in Section 2.1 and also consider an extension, the *Riemann–Stieltjes integral*. The latter can be applied to Brownian sample paths if the integrand process is sufficiently smooth. However, we will see in Section 2.1.2 that *Brownian sample paths cannot be integrated with respect to themselves*, which is a major handicap. Therefore we are forced to consider an integral which is not defined path by path. This will be done in Section 2.2. There we define the *Itô stochastic integral* as a mean square limit of certain Riemann–Stieltjes sums. This has the disadvantage that we lose the intuitive interpretation of such an integral which is naturally provided by a pathwise integral. However, it will be convenient to think of the stochastic integral in terms of an approximating Riemann–Stieltjes sum.

In Section 2.3 we will learn about another important tool: the *Itô lemma*. It is the stochastic analogue of the ordinary chain rule of differentiation. The Itô lemma will be crucial in Chapter 3, where we intend to solve Itô stochastic differential equations.

In Section 2.4 we will learn about one more stochastic integral whose value usually differs from the Itô stochastic integral. It is called the *Stratonovich stochastic integral* and can be useful as an auxiliary tool for the solution of Itô stochastic differential equations.

CHAPTER 2.

2.1 The Riemann and Riemann–Stieltjes Integrals

This section is not needed for the understanding of the Itô stochastic integral, but it gives some information about the history of integration and explains why classical integrals may fail when the integrand or the integrator are Brownian sample paths.

We discuss some approaches to pathwise integration. In particular, we are interested in integrals of the form $\int_0^t f(s)\,dB_s(\omega)$, where $(B_t(\omega), t \geq 0)$ is a given Brownian sample path and f is a deterministic function or the sample path of a stochastic process. In particular, we recall the Riemann integral $\int_0^1 f(t)\,dt$ as *the* classical integral, and some of its properties. Then we discuss Riemann–Stieltjes integrals which are close in spirit to the pathwise integral $\int_0^1 f(t)\,dB_t(\omega)$. We will see that, under some smoothness conditions on f, the latter integral is well defined, and we will also point out that, for example, the Riemann–Stieltjes integral $\int_0^1 B_t(\omega)\,dB_t(\omega)$ cannot be defined.

2.1.1 The Ordinary Riemann Integral

We want to recall the notion of *ordinary integral*, which is also called the *Riemann integral*. You should be familiar with this integral from a course on elementary calculus.

Suppose, for simplicity only, that f is a real-valued function defined on $[0, 1]$, but instead we could consider any interval $[a, b]$.

Consider a *partition* of the interval $[0, 1]$:

$$\tau_n: \quad 0 = t_0 < t_1 < \cdots < t_{n-1} < t_n = 1\,,$$

and define

$$\Delta_i = t_i - t_{i-1}\,, \quad i = 1,\dots,n\,.$$

An *intermediate partition* σ_n of τ_n is given by any values y_i satisfying $t_{i-1} \leq y_i \leq t_i$ for $i = 1,\dots,n$. For given partitions τ_n and σ_n we can define the *Riemann sum*

$$S_n = S_n(\tau_n, \sigma_n) = \sum_{i=1}^{n} f(y_i)\,(t_i - t_{i-1}) = \sum_{i=1}^{n} f(y_i)\,\Delta_i\,. \qquad (2.1)$$

Thus a Riemann sum is nothing but a weighted average of the values $f(y_i)$, where the weights are the corresponding lengths Δ_i of the intervals $[t_{i-1}, t_i]$. We also know that S_n is an approximation to the area between the graph of

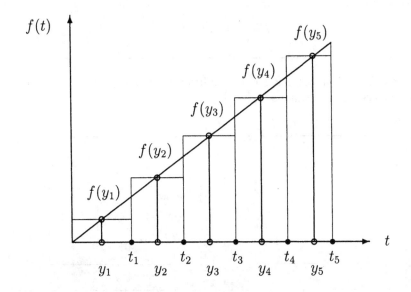

Figure 2.1.1 *An illustration of a Riemann sum with partition (t_i) and intermediate partition (y_i). The sum of the rectangular areas approximates the area between the graph of $f(t)$ and the t-axis, i.e. $\int_0^{t_5} f(t)\,dt$.*

the function f and the t-axis, provided f only assumes non-negative values. See Figure 2.1.1 for an illustration.

Now let the mesh of the partition τ_n go to zero, i.e.

$$\mathrm{mesh}(\tau_n) := \max_{i=1,\ldots,n} \Delta_i = \max_{i=1,\ldots,n} (t_i - t_{i-1}) \to 0.$$

If we proceed in this way, the points $t_i = t_i^{(n)}$ clearly have to depend on n, but we suppress this dependence in our notation.

If the limit

$$S = \lim_{n \to \infty} S_n = \lim_{n \to \infty} \sum_{i=1}^{n} f(y_i)\,\Delta_i$$

exists as $\mathrm{mesh}(\tau_n) \to 0$ and S is independent of the choice of the partitions τ_n and their intermediate partitions σ_n, then S is called the *ordinary* or *Riemann integral* of f on $[0, 1]$.

We write

$$S = \int_0^1 f(t)\, dt\,.$$

We know that $\int_0^1 f(t)\, dt$ exists if f is sufficiently smooth, for example, continuous or piecewise continuous on $[0,1]$.

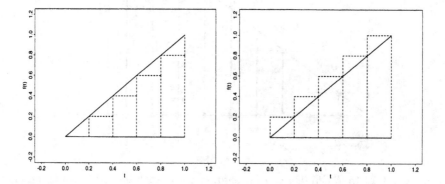

Figure 2.1.2 *Two Riemann sums for the integral $\int_0^1 t\, dt$; see Example 2.1.3. The partition is given by $t_i = i/5$, $i = 0, \ldots, 5$. The left (left figure) and the right (right figure) end points of the intervals $[(i-1)/5, i/5]$ are taken as points y_i of the intermediate partition.*

Example 2.1.3 We want to calculate the integral $\int_0^1 t\, dt$ as the limit of certain Riemann sums. See Figure 2.1.2 for an illustration. We already know that $\int_0^1 t\, dt = 0.5$; it represents the triangular area between the graph of $f(t) = t$ and the t-axis. For numerical approximations it is convenient to choose an equidistant partition of $[0,1]$:

$$t_i = \frac{i}{n}\,, \quad i = 0, \ldots, n\,.$$

First take the left end points of the intervals $[(i-1)/n, i/n]$ for the intermediate partition:

$$S_n^{(l)} = \sum_{i=1}^n \frac{i-1}{n} \Delta_i = \frac{1}{n} \sum_{i=1}^n \frac{i-1}{n}\,.$$

Using the well-known sum formula

$$\sum_{i=1}^{n} i = \frac{n(n+1)}{2} ,$$ (2.2)

we conclude that

$$S_n^{(l)} = \frac{1}{n^2} \frac{n(n-1)}{2} \ \to \ \frac{1}{2} = S .$$

Analogously, take the right end points of the intervals $[(i-1)/n, i/n]$:

$$S_n^{(r)} = \sum_{i=1}^{n} \frac{i}{n} \Delta_i = \frac{1}{n} \sum_{i=1}^{n} \frac{i}{n} ,$$

and conclude, again using (2.2), that

$$S_n^{(r)} = \frac{1}{n^2} \frac{n(n+1)}{2} \ \to \ \frac{1}{2} = S .$$

Now choose the y_is as the middle points of the intervals $[(i-1)/n, i/n]$ and denote the corresponding Riemann sums by $S_n^{(m)}$. Since $S_n^{(l)} \le S_n^{(m)} \le S_n^{(r)}$ we may conclude that $S_n^{(m)} \to S = 0.5$. □

The Riemann integral is taken as a model for the definition of any type of integral. The new kind of integral should have as many properties in common with the Riemann integral as possible. Such properties are given below.

For Riemann integrable functions f, f_1 and f_2 on $[0,1]$ the following properties hold:

- The Riemann integral is *linear*, i.e. for any constants c_1 and c_2 :

$$\int_0^1 [c_1 \, f_1(t) + c_2 \, f_2(t)] \, dt = c_1 \int_0^1 f_1(t) \, dt + c_2 \int_0^1 f_2(t) \, dt .$$

- The Riemann integral is linear on adjacent intervals:

$$\int_0^1 f(t) \, dt = \int_0^a f(t) \, dt + \int_a^1 f(t) \, dt \quad \text{for} \quad 0 \le a \le 1 .$$

- One can define the *indefinite Riemann integral* as a function of the upper limit:

$$\int_0^s f(t)\, dt = \int_0^1 f(t)\, I_{[0,s]}(t)\, dt\,, \quad 0 \le s \le 1\,.$$

2.1.2 The Riemann–Stieltjes Integral

In probability theory it is usual to denote the expectation of a random variable X by

$$EX = \int_{-\infty}^{\infty} t\, dF_X(t)\,,$$

where F_X denotes the distribution function of X. This notation refers to the fact that EX is defined as a Riemann–Stieltjes integral, or as a Lebesgue–Stieltjes integral. This means, roughly speaking, that

$$\int_{-\infty}^{\infty} t\, dF_X(t) \approx \sum_i y_i\, [F_X(t_i) - F_X(t_{i-1})]$$

for a partition (t_i) of $\mathbb{R}$ and a corresponding intermediate partition (y_i). Also recall from a course on elementary calculus that you can define the integral $\int_0^1 f(t)\, dg(t)$ as $\int_0^1 f(t) g'(t)\, dt$, provided the derivative $g'(t)$ exists.

Both, $\int_{-\infty}^{\infty} t\, dF_X(t)$ and $\int_0^1 f(t)\, dg(t)$, are examples of how one could approach the problem of integrating one function f with respect to another function g. It is the aim of the present chapter to suggest some methods. In particular, we would like to obtain an integral of type $\int_0^1 f(t)\, dB_t(\omega)$, where f is a function or a stochastic process on $[0,1]$ and $B_t(\omega)$ is a Brownian sample path. We have a major difficulty in defining such an integral since the path $B_t(\omega)$ does not have a derivative; see Section 1.3.1. However, one can find some kind of pathwise integral which sometimes allows one to evaluate the integral $\int_0^1 f(t)\, dB_t(\omega)$. This kind of integral is called the *Riemann–Stieltjes integral*. It is a classical tool in many fields of mathematics.

In what follows we give a precise definition of the Riemann–Stieltjes integral. As for the Riemann integral, we consider a partition of the interval $[0,1]$:

$$\tau_n : \quad 0 = t_0 < t_1 < \cdots < t_{n-1} < t_n = 1\,,$$

and an *intermediate partition* σ_n *of* τ_n:

$$\sigma_n : \quad t_{i-1} \leq y_i \leq t_i \quad \text{for} \quad i = 1, \ldots, n.$$

Let f and g be two real-valued functions on $[0, 1]$ and define

$$\Delta_i g = g(t_i) - g(t_{i-1}), \quad i = 1, \ldots, n.$$

The *Riemann–Stieltjes sum* corresponding to τ_n and σ_n is given by

$$S_n = S_n(\tau_n, \sigma_n) = \sum_{i=1}^{n} f(y_i) \, \Delta_i g = \sum_{i=1}^{n} f(y_i) \, [g(t_i) - g(t_{i-1})].$$

Notice the similarity with a Riemann sum; see (2.1). In that case, $g(t) = t$. Thus, a Riemann–Stieltjes sum is obtained by weighing the values $f(y_i)$ with the increments $\Delta_i g$ of g in the intervals $[t_{i-1}, t_i]$.

If the limit

$$S = \lim_{n \to \infty} S_n = \lim_{n \to \infty} \sum_{i=1}^{n} f(y_i) \, \Delta_i g$$

exists as $\text{mesh}(\tau_n) \to 0$ and S is independent of the choice of the partitions τ_n and their intermediate partitions σ_n, then S is called the *Riemann–Stieltjes integral of f with respect to g on $[0, 1]$.*
We write

$$S = \int_0^1 f(t) \, dg(t).$$

It raises the question:

When does the Riemann-Stieltjes integral $\int_0^1 f(t) \, dg(t)$ exist, and is it possible to take $g = B$ for Brownian motion B on $[0, 1]$?

This question does not have a simple answer and needs some more sophistication. In some textbooks one can find the following partial answers:

- The functions f and g must not have discontinuities at the same point $t \in [0, 1]$.

- Assume that f is continuous and g has *bounded variation*, i.e.

$$\sup_{\tau} \sum_{i=1}^{n} |g(t_i) - g(t_{i-1})| < \infty,$$

where the supremum (cf. its definition on p. 211) is taken over all possible partitions τ of $[0, 1]$.

Then the Riemann–Stieltjes integral $\int_0^1 f(t)\, dg(t)$ exists.

Notice that the latter result is not applicable to the integral $\int_0^1 f(t)\, dB_t(\omega)$. Indeed, we learnt in Section 1.3.1 that Brownian sample paths $B_t(\omega)$ do not have bounded variation. However,

> *bounded variation of g is* not *necessary for the existence of the Riemann–Stieltjes integral $\int_0^1 f(t)\, dg(t)$,*

although the contrary claim can be found in some books. Weaker conditions for the existence of $\int_0^1 f(t)\, dg(t)$ are not very well known, but they were already found by L.C. Young in 1936; see the recent papers by Dudley and Norvaiša (1998a,b) for an extensive discussion.

Without going into detail, we give a sufficient condition for Riemann–Stieltjes integrability which is close to necessity. First we give a definition:

The real-valued function h on $[0, 1]$ is said to have *bounded p-variation* for some $p > 0$ if

$$\sup_\tau \sum_{i=1}^n |h(t_i) - h(t_{i-1})|^p < \infty,$$

where the supremum is taken over all partitions τ of $[0, 1]$.

Notice that h has *bounded variation* if $p = 1$.

The Riemann–Stieltjes integral $\int_0^1 f(t)\, dg(t)$ exists if the following conditions are satisfied:

- The functions f and g do not have discontinuities at the same point $t \in [0, 1]$.

- The function f has bounded p-variation and the function g has bounded q-variation for some $p > 0$ and $q > 0$ such that $p^{-1} + q^{-1} > 1$.

In some cases these conditions can be verified. We give an example for the integral $\int_0^1 f(t)\, dB_t(\omega)$.

Example 2.1.4 As before, $B = (B_t, t \geq 0)$ denotes Brownian motion. It is well known that sample paths of Brownian motion have bounded p-variation on any fixed finite interval, provided that $p > 2$, and unbounded p-variation for $p \leq 2$. See Taylor (1972).

Now consider a deterministic function $f(t)$ on $[0, 1]$ or a sample path of a stochastic process $f(t, \omega)$. According to the above theory, we can define the Riemann–Stieltjes integral $\int_0^1 f(t) \, dB_t(\omega)$ with respect to the sample path $B_t(\omega)$, provided f has bounded q-variation for some $q < 2$. This is satisfied if f has bounded variation, i.e. $q = 1$.

Assume that f is a differentiable function with bounded derivative $f'(t)$. Then it follows by an application of the mean value theorem that

$$|f(t) - f(s)| \leq K(t - s), \quad s < t,$$

where $K > 0$ is a constant depending on f. Then

$$\sup_\tau \sum_{i=1}^n |f(t_i) - f(t_{i-1})| \leq K \sum_{i=1}^n (t_i - t_{i-1}) = K < \infty.$$

Hence f has bounded variation, and so the following statement holds:

Assume f is a deterministic function or the sample path of a stochastic process. If f is differentiable with a bounded derivative on $[0, 1]$, then the Riemann–Stieltjes integral

$$\int_0^1 f(t) \, dB_t(\omega)$$

exists for every Brownian sample path $B_t(\omega)$.

In particular, you can define the following integrals as Riemann–Stieltjes integrals:

$$\int_0^1 e^t \, dB_t(\omega), \quad \int_0^1 \sin(t) \, dB_t(\omega), \quad \int_0^1 t^p \, dB_t(\omega), \quad p \geq 0. \tag{2.3}$$

This does not mean that you can evaluate these integrals explicitly in terms of Brownian motion. $\quad\square$

The above discussion may be slightly misleading since it suggests that one can define the integral $\int_0^1 f(t) \, dB_t(\omega)$ for very general integrands f as a Riemann–Stieltjes integral with respect to a Brownian sample path. However, this is not

the case. A famous example, which is also one of the motivating examples for
the introduction of the Itô stochastic integral, is the integral

$$I(B)(\omega) = \int_0^1 B_t(\omega) \, dB_t(\omega) \, .$$

Brownian motion has bounded p-variation for $p > 2$, not for $p \leq 2$, and so
the sufficient condition $2p^{-1} > 1$ for the existence of $I(B)$ (see p. 94) is not
satisfied. Furthermore, it can be shown that this simple integral does not exist
as a Riemann–Stieltjes integral.

Moreover, one can show the following: if $\int_0^1 f(t) \, dg(t)$ exists as a Riemann–
Stieltjes integral for *all* continuous functions f on $[0, 1]$, then g necessarily has
bounded variation; see Protter (1992), Theorem 52. Since one of our aims is
to define the integrals $\int_0^1 f(t) \, dB_t(\omega)$ for all continuous deterministic functions
f on $[0, 1]$, the Riemann–Stieltjes integral approach must fail since Brownian
sample paths do not have bounded variation on any finite interval.

It is the objective of the following sections to find another approach to
the stochastic integral. Since pathwise integration with respect to a Brownian
sample path, as suggested by the Riemann–Stieltjes integral, does not lead to
a sufficiently large class of integrable functions f, we will try to define the
integral as a probabilistic average. This approach has the disadvantage in that
it is less intuitive than the Riemann–Stieltjes integral, regarding the form of
the integrals.

Notes and Comments

The Riemann and Riemann–Stieltjes integrals are treated in textbooks on
calculus. However, it is not easy to find a comprehensive treatment of the
Riemann–Stieltjes integral. Young (1936) is still one of the best references
on the topic. Extensions of the Riemann–Stieltjes integral are treated in the
recent work by Dudley and Norvaiša (1998a,b).

2.2   The Itô Integral

2.2.1   A Motivating Example

At the end of the previous section we learnt that it is impossible to define the
integral $\int_0^1 B_t(\omega) \, dB_t(\omega)$ path by path as a Riemann–Stieltjes integral. Here
and in what follows, $B = (B_t, t \geq 0)$ denotes Brownian motion. In order to

understand what is going on, we consider the Riemann–Stieltjes sums

$$S_n = \sum_{i=1}^{n} B_{t_{i-1}} \Delta_i B, \tag{2.4}$$

where

$$\tau_n : \quad 0 = t_0 < t_1 < \cdots < t_{n-1} < t_n = t, \tag{2.5}$$

is a partition of $[0, t]$ and, for any function f on $[0, t]$,

$$\Delta f : \quad \Delta_i f = f(t_i) - f(t_{i-1}), \quad i = 1, \ldots, n,$$

are the corresponding increments of f and

$$\Delta_i = t_i - t_{i-1}, \quad i = 1, \ldots, n.$$

Thus the Riemann–Stieltjes sum S_n corresponds to the partition τ_n and the intermediate partition (y_i) with $y_i = t_{i-1}$, i.e. y_i is the left end point of the interval $[t_{i-1}, t_i]$. This choice is typical for the definition of the Itô stochastic integral. We will also see that another choice of intermediate partition (y_i) gives rise to the definition of another type of stochastic integral.

Notice that S_n can be written in the following form:

$$S_n = \frac{1}{2} B_t^2 - \frac{1}{2} \sum_{i=1}^{n} (\Delta_i B)^2 =: \frac{1}{2} B_t^2 - \frac{1}{2} Q_n(t).$$

(Check it by using the binomial formula

$$(B_{t_i} - B_{t_{i-1}})^2 = B_{t_i}^2 + B_{t_{i-1}}^2 - 2 B_{t_i} B_{t_{i-1}}.)$$

Recall that Brownian motion has independent and stationary increments (see p. 33). Hence

$$E(\Delta_i B \, \Delta_j B) = \begin{cases} 0, & \text{if } i \neq j, \\ \text{var}(\Delta_i B) = t_i - t_{i-1} = \Delta_i, & \text{if } i = j. \end{cases}$$

Then it is immediate that

$$E Q_n(t) = \sum_{i=1}^{n} E(\Delta_i B)^2 = \sum_{i=1}^{n} \Delta_i = t,$$

and, again by the independent increments of Brownian motion and since the relation $\text{var}(X) = E(X^2) - (EX)^2$ holds,

$$\text{var}(Q_n(t)) \quad = \quad \sum_{i=1}^{n} \text{var}([\Delta_i B]^2) = \sum_{i=1}^{n} [E(\Delta_i B)^4 - \Delta_i^2].$$

It is a well-known fact that for the $N(0,1)$ random variable B_1, $EB_1^4 = 3$. Hence,

$$E(\Delta_i B)^4 = EB_{t_i - t_{i-1}}^4 = E\left[(\Delta_i)^{1/2} B_1\right]^4 = 3\Delta_i^2,$$

which implies that

$$\text{var}(Q_n(t)) \quad = \quad 2 \sum_{i=1}^{n} \Delta_i^2.$$

Thus, if

$$\text{mesh}(\tau_n) \quad = \quad \max_{i=1,\ldots,n} \Delta_i \to 0,$$

we obtain that

$$\text{var}(Q_n(t)) \leq 2\,\text{mesh}(\tau_n) \sum_{i=1}^{n} \Delta_i = 2t\,\text{mesh}(\tau_n) \to 0.$$

Since $\text{var}(Q_n(t)) = E(Q_n(t) - t)^2$ we showed:

$Q_n(t)$ converges to t in mean square, hence in probability; see Appendix A1 for the different modes of convergence.

The limiting function $f(t) = t$ is characteristic only for Brownian motion. It is called *the quadratic variation of Brownian motion on* $[0, t]$.

One can show that $Q_n(t)$ does not converge for a given Brownian sample path and suitable choices of partitions τ_n. This fact clearly indicates that we cannot define $\int_0^t B_s(\omega)\,dB_s(\omega)$ as a Riemann–Stieltjes integral. However,

we could define $\int_0^t B_s\,dB_s$ as a mean square limit.

Indeed, since $S_n = 0.5\,[B_t^2 - Q_n(t)]$ converges in mean square to $0.5\,(B_t^2 - t)$, we could take this limit as the value of the integral $\int_0^t B_s\,dB_s$. Later it will turn out that this is the value of the Itô stochastic integral:

$$\int_0^t B_s\,dB_s \quad = \quad \frac{1}{2}\,(B_t^2 - t). \qquad (2.6)$$

From this evaluation we have learnt various facts which will be useful in what follows:

> Integrals with respect to Brownian sample paths, which cannot be defined in the Riemann–Stieltjes sense, can hopefully be defined in the mean square sense.

The increment $\Delta_i B = B_{t_i} - B_{t_{i-1}}$ on the interval $[t_{i-1}, t_i]$ satisfies $E\Delta_i B = 0$ and $E(\Delta_i B)^2 = \Delta_i = t_i - t_{i-1}$. The mean square limit of $Q_n(t)$ is t. These properties suggest that $(\Delta_i B)^2$ is of the order Δ_i.

> In terms of differentials, we write
>
> $$(dB_t)^2 = (B_{t+dt} - B_t)^2 = dt\,, \qquad (2.7)$$
>
> and in terms of integrals,
>
> $$\int_0^t (dB_s)^2 = \int_0^t ds = t\,. \qquad (2.8)$$
>
> The right-hand side is the quadratic variation of Brownian motion on $[0, t]$.

At the moment, relations (2.7) and (2.8) are nothing but heuristic rules. They can be made mathematically correct (in the mean square sense). In this book we do not have the mathematical means to prove them. In what follows, we will use (2.7) and (2.8) as given rules; they will help us to understand the results of Itô calculus, in particular the Itô lemma in Section 2.3.

Next we try to understand why we chose the Riemann–Stieltjes sums S_n in (2.4) in such a specific way: the values of Brownian motion were evaluated at the left end points of the intervals $[t_{i-1}, t_i]$. Assume that we have a partition τ_n of $[0, t]$ as in (2.5). We learnt in Example 1.5.7 that the martingale transform $\widetilde{B} \bullet \Delta B$ given by

$$\sum_{i=1}^k \widetilde{B}_i \, \Delta_i B = \sum_{i=1}^k B_{t_{i-1}} (B_{t_i} - B_{t_{i-1}})\,, \quad k = 1, \ldots, n\,,$$

is a martingale with respect to the filtration $\sigma(B_{t_i}, i = 1, \ldots, k)$, $k = 1, \ldots, n$. As a result, the mean square limit $0.5\,(B_t^2 - t)$ of the approximating Riemann–Stieltjes sums is a martingale with respect to the natural Brownian filtration.

Another definition of the Riemann–Stieltjes sums makes them lose the martingale property. As a matter of fact we mention that, for partitions $\tau_n = (t_i)$ of $[0, t]$ with mesh$(\tau_n) \to 0$, the Riemann–Stieltjes sums

$$\widetilde{S}_n = \sum_{i=1}^{n} B_{y_i}\, \Delta_i B$$

with

$$y_i = \frac{1}{2}\left(t_{i-1} + t_i\right), \quad i = 1, \ldots, n,$$

have the mean square limit $0.5\, B_t^2$ (you can check this by the same arguments and tools as given for S_n). This quantity can be interpreted as the value of another stochastic integral, $\int_0^t B_s \circ dB_s$ say. The Riemann–Stieltjes sums $\sum_{i=1}^{k} B_{y_i}\, \Delta_i B$, $k = 1, \ldots, n$, do *not* constitute a martingale, and neither does the limit process $0.5 B_t^2$ (check it!). However, the latter process enjoys another nice property. The relation

$$\int_0^t B_s \circ dB_s = \frac{1}{2} B_t^2 \tag{2.9}$$

indicates that the *classical chain rule* holds.

To be precise, let $b(t)$ be a deterministic differentiable function with $b(0) = 0$. We know that

$$\frac{1}{2}\frac{db^2(s)}{ds} = b(s)\frac{db(s)}{ds}\,.$$

Integrating both sides, we obtain by the classical rules of calculus,

$$\frac{1}{2}\int_0^t \frac{db^2(s)}{ds}\, ds = \frac{1}{2} b^2(t) = \int_0^t b(s)\frac{db(s)}{ds}\, ds = \int_0^t b(s)\, db(s)\,.$$

If we *formally* replace the function $b(t)$ with Brownian motion B_t, we obtain the same value as for the stochastic integral (2.9). However, it is only a *formal* replacement, since this chain rule is applicable to differentiable functions, not to Brownian sample paths.

The stochastic integral, which is obtained as the mean square limit of the Riemann–Stieltjes sums, evaluated at the middle points of the intervals $[t_{i-1}, t_i]$, is called a *Stratonovich integral*. We will consider it in Section 2.4. It will turn out to be a useful tool for solving Itô stochastic differential equations.

Thus we have learnt one more useful fact:

The formula $\int_0^t B_s\, dB_s = 0.5\,(B_t^2 - t)$ suggests that the classical chain rule of integration does not hold for Itô stochastic integrals. In Section 2.3 we will find a chain rule which is well suited for Itô integration: the *Itô lemma*.

2.2.2 The Itô Stochastic Integral for Simple Processes

We start the investigation of the Itô stochastic integral for a class of processes whose paths assume only a finite number of values. As usual, $B = (B_t, t \geq 0)$ denotes Brownian motion, and

$$\mathcal{F}_t = \sigma(B_s\,,\, s \leq t)\,, \quad t \geq 0\,,$$

is the corresponding natural filtration. Recall that a stochastic process $X = (X_t\,,\, t \geq 0)$ is adapted to Brownian motion if X is adapted to $(\mathcal{F}_t, t \geq 0)$. This means that, for every t, X_t is a function of the past and present of Brownian motion.

In what follows, we consider all processes on a fixed interval $[0, T]$.

First we introduce an appropriate class of Itô integrable processes.

The stochastic process $C = (C_t, t \in [0, T])$ is said to be *simple* if it satisfies the following properties:

There exists a partition

$$\tau_n : \quad 0 = t_0 < t_1 < \cdots < t_{n-1} < t_n = T\,,$$

and a sequence $(Z_i, i = 1, \ldots, n)$ of random variables such that

$$C_t = \begin{cases} Z_n\,, & \text{if } t = T\,, \\ Z_i\,, & \text{if } t_{i-1} \leq t < t_i\,, \quad i = 1, \ldots, n\,. \end{cases}$$

The sequence (Z_i) is adapted to $(\mathcal{F}_{t_{i-1}}, i = 1, \ldots, n)$, i.e. Z_i is a function of Brownian motion up to time t_{i-1}, and satisfies $E Z_i^2 < \infty$ for all i.

Example 2.2.1 (Some simple processes)
The deterministic function

$$f_n(t) = \begin{cases} \dfrac{n-1}{n}\,, & \text{if } t = T\,, \\[2mm] \dfrac{i-1}{n}\,, & \text{if } \dfrac{i-1}{n} \leq t < \dfrac{i}{n}\,, \quad i = 1, \ldots, n\,, \end{cases}$$

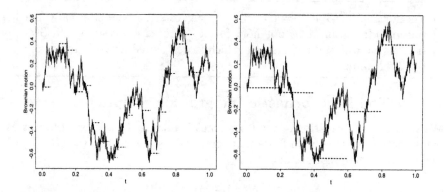

Figure 2.2.2 *Two approximations of a Brownian sample path by simple processes C (dashed lines), given by (2.10).*

is a step function, hence it is a simple process.

Now define the process

$$
C_t = \begin{cases} Z_n = B_{t_{n-1}}, & \text{if } t = T, \\ Z_i = B_{t_{i-1}}, & \text{if } t_{i-1} \le t < t_i, \quad i = 1, \dots, n, \end{cases} \tag{2.10}
$$

for a given partition τ_n of $[0, T]$. It is a simple process: the paths are piecewise constant, and C_t is a function of Brownian motion until time t. For an illustration of the process C for two different partitions, see Figure 2.2.2.   □

The *Itô stochastic integral of a simple process C* on $[0, T]$ is given by

$$
\int_0^T C_s \, dB_s := \sum_{i=1}^n C_{t_{i-1}} (B_{t_i} - B_{t_{i-1}}) = \sum_{i=1}^n Z_i \, \Delta_i B.
$$

The *Itô stochastic integral of a simple process* C on $[0,t]$, $t_{k-1} \leq t \leq t_k$, is given by

$$\int_0^t C_s \, dB_s :=$$

$$\int_0^T C_s \, I_{[0,t]}(s) \, dB_s = \sum_{i=1}^{k-1} Z_i \, \Delta_i B + Z_k \, (B_t - B_{t_{k-1}}), \quad (2.11)$$

where $\sum_{i=1}^0 Z_i \, \Delta_i B = 0$.

Thus the value of the Itô stochastic integral $\int_0^t C_s \, dB_s$ is the Riemann–Stieltjes sum of the path of C, evaluated at the left end points of the intervals $[t_{i-1}, t_i]$, with respect to Brownian motion. If $t < t_n$, the point t can formally be interpreted as the last point of the partition of $[0, t]$.

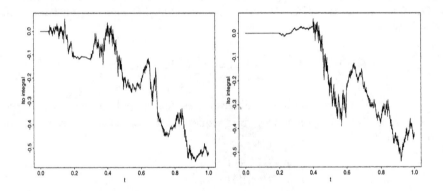

Figure 2.2.3 *The Itô stochastic integral $\int_0^t C_s \, dB_s$ corresponding to the paths of C and B given in Figure 2.2.2.*

Example 2.2.4 (Continuation of Example 2.2.1)
Recall the simple processes f_n and C from Example 2.2.1. The corresponding Itô stochastic integrals are then given by

$$\int_0^t f_n(s) \, dB_s = \sum_{i=1}^{k-1} \frac{i-1}{n} \left(B_{i/n} - B_{(i-1)/n} \right) + \frac{k-1}{n} \left(B_t - B_{(k-1)/n} \right),$$

for $t \in [(k-1)/n, k/n]$, and by

$$\int_0^t C_s \, dB_s = \sum_{i=1}^{k-1} B_{t_{i-1}} \Delta_i B + B_{t_{k-1}} \left(B_t - B_{t_{k-1}} \right), \qquad (2.12)$$

for $t \in [t_{k-1}, t_k]$. See Figures 2.2.2 and 2.2.3 for a visualization of the sample paths of B, C and the corresponding Itô stochastic integrals.

We know from Example 2.1.4 that

$$\lim_{n \to \infty} \int_0^t f_n(s) \, dB_s(\omega) = \int_0^t s \, dB_s(\omega),$$

where the right-hand side is a Riemann–Stieltjes integral with respect to a given Brownian sample path. We also learnt in Section 2.1.2 that the Riemann–Stieltjes sums (2.12) do not in general converge, as mesh(τ_n) $\to 0$, for a given sample path of Brownian motion. $\qquad \qquad \qquad \qquad \qquad \qquad \qquad \qquad \square$

The form of the Itô stochastic integral for simple processes very much reminds us of a martingale transform. On p. 83 we introduced the martingale transform $\widetilde{C} \bullet Y$, where

$$(\widetilde{C} \bullet Y)_0 = 0, \quad (\widetilde{C} \bullet Y)_n = \sum_{i=1}^n \widetilde{C}_i Y_i, \quad n = 1, 2, \dots.$$

Here $Y = (Y_n)$ is a martingale difference sequence with respect to a given filtration and $\widetilde{C} = (\widetilde{C}_n)$ is a previsible sequence. In this sense, the sequence of Itô stochastic integrals

$$\left(\int_0^{t_k} C_s \, dB_s, \quad k = 0, \dots, n \right)$$

of a simple process $(C_s, s \leq t)$ is a martingale transform with respect to the filtration $(\mathcal{F}_{t_k}, k = 0, \dots, n)$.

But even more is true:

> The stochastic process $I_t(C) = \int_0^t C_s \, dB_s$, $t \in [0, T]$, is a martingale with respect to the natural Brownian filtration $(\mathcal{F}_t, t \in [0, T])$.

We check the defining properties of a martingale; see p. 80. For convenience we recall them in terms of $I(C)$:

• $E|I_t(C)| < \infty$ for all $t \in [0, T]$.

- $I(C)$ is adapted to $(\mathcal{F}_t)$.

-
$$E(I_t(C) \mid \mathcal{F}_s) = I_s(C) \quad \text{for} \quad s < t. \tag{2.13}$$

The first property follows, for example, from the isometry property (2.14) given below.

Adaptedness of $I(C)$ to Brownian motion is easily seen: at time t, the random variables $Z_1, \ldots, Z_k$ and $\Delta_1 B, \ldots, \Delta_{k-1} B$, $B_t - B_{t_{k-1}}$, occurring in the defining relation (2.11), are all functions of Brownian motion up to time t.

It remains to show the crucial relation (2.13). This is a good exercise in using the rules of conditional expectation, which we suggest you try for yourself. We have not indicated which rules to use as we assume that you know them by now.

First assume that $s < t$ and $s, t \in [t_{k-1}, t_k]$. Notice that

$$
\begin{aligned}
I_t(C) &= I_{t_{k-1}}(C) + Z_k (B_s - B_{t_{k-1}}) + Z_k (B_t - B_s) \\
&= I_s(C) + Z_k (B_t - B_s),
\end{aligned}
$$

where $I_s(C)$ and Z_k are functions of Brownian motion up to time s, and $B_t - B_s$ is independent of $\mathcal{F}_s$. Hence,

$$E(I_t(C) \mid \mathcal{F}_s) = I_s(C) + Z_k E(B_t - B_s) = I_s(C).$$

This proves (2.13) in this case.

The case, when $s \in [t_{l-1}, t_l]$ and $t \in [t_{k-1}, t_k]$ for some $l < k$, can be handled analogously. Check (2.13); make use of the decomposition

$$
\begin{aligned}
I_t(C) &= \left[I_{t_{l-1}}(C) + Z_l (B_s - B_{t_{l-1}}) \right] \\
&\quad + \left[Z_l (B_{t_l} - B_s) + \sum_{i=l+1}^{k-1} Z_i \Delta_i B + Z_k (B_t - B_{t_{k-1}}) \right] \\
&= I_s(C) + R(s,t),
\end{aligned}
$$

and show that $E(R(s,t) \mid \mathcal{F}_s) = 0$.

Properties of the Itô Stochastic Integral for Simple Processes

It is easy to see that $E I_t(C) = 0$ since Z_i and $\Delta_i B$ in the definition (2.11) are independent, and $E(Z_i \Delta_i B) = E Z_i E \Delta_i B = 0$. Thus:

> The Itô stochastic integral has expectation zero.

Alternatively, we can argue as follows: since $I(C)$ is a martingale, it has a constant expectation function. Moreover, by definition, $I_0(C) = 0$, and so $EI_t(C) = EI_0(C) = 0$.

Another property of the Itô stochastic integral for simple processes will turn out to be crucial for the definition of the general Itô stochastic integral.

> The Itô stochastic integral satisfies the *isometry property* :
>
> $$ E\left(\int_0^t C_s\, dB_s\right)^2 = \int_0^t EC_s^2\, ds\,, \quad t \in [0, T]\,. \qquad (2.14) $$

We show this property. For the ease of presentation, we assume that $t = t_k$ for some k. Indeed, if $t_{k-1} < t < t_k$, observe that $0 = t_0 < \cdots < t_{k-1} < t'_k := t$ is a partition of $[0, t]$, so you can formally consider t as a point of the partition. We have, for $W_i = Z_i\, \Delta_i B$,

$$ E[I_t(C)]^2 = \sum_{i=1}^{k} \sum_{j=1}^{k} E(W_i W_j)\,. \qquad (2.15) $$

You can check that, for $i > j$, the random variables W_i and W_j are uncorrelated: notice that W_j and Z_i are functions of Brownian motion up to time t_{i-1}, hence they are independent of $\Delta_i B$. We conclude that

$$ E(W_i W_j) = E(W_j Z_i)\, E(\Delta_i B) = 0\,. $$

Thus the terms $E(W_i W_j)$ in (2.15) vanish for $i \neq j$, and so we obtain

$$ E[I_t(C)]^2 = \sum_{i=1}^{k} E(Z_i\, \Delta_i B)^2 = \sum_{i=1}^{k} EZ_i^2\, E(\Delta_i B)^2 = \sum_{i=1}^{k} EZ_i^2\, (t_i - t_{i-1})\,. $$

The right-hand side is nothing but the Riemann integral $\int_0^t f(s)\, ds$ of the step function $f(s) = EC_s^2$, which coincides with EZ_i^2 for $t_{i-1} \leq t < t_i$ (check this!). Thus we have proved the isometry property of the Itô stochastic integral.

The Itô stochastic integral has several properties in common with the Riemann and Riemann–Stieltjes integrals.

The Itô stochastic integral is *linear:*

For constants c_1, c_2 and simple processes $C^{(1)}$ and $C^{(2)}$ on $[0, T]$,

$$\int_0^t \left[c_1\, C_s^{(1)} + c_2\, C_s^{(2)} \right] dB_s = c_1 \int_0^t C_s^{(1)}\, dB_s + c_2 \int_0^t C_s^{(2)}\, dB_s \,.$$

The Itô stochastic integral is *linear on adjacent intervals:*

For $0 \le t \le T$,

$$\int_0^T C_s\, dB_s = \int_0^t C_s\, dB_s + \int_t^T C_s\, dB_s \,.$$

The proof of the linearity is not difficult; try it yourself. Before you start you have to define $C^{(1)}$ and $C^{(2)}$ on the same partition. This is always possible: if $\tau_n^{(1)}$ is the partition corresponding to $C^{(1)}$ and $\tau_m^{(2)}$ the partition corresponding to $C^{(2)}$, you get a joint partition τ with at most $n+m$ distinct points by taking the union of $\tau_n^{(1)}$ and $\tau_m^{(2)}$. Clearly, this is a refinement of the two original partitions. The values of the $I(C^{(i)})$s remain the same with this new partition. Linearity follows from the linearity of the underlying Riemann–Stieltjes sums.

The linearity on adjacent intervals follows from linearity since

$$\int_0^T C_s\, dB_s = \int_0^T \left[C_s^{(1)} + C_s^{(2)} \right] dB_s \,,$$

where $C_s^{(1)} = C_s I_{[0,t]}(s)$ and $C_s^{(2)} = C_s I_{(t,T]}(s)$ are two simple processes.

Finally, we state the following property:

The process $I(C)$ has continuous sample paths.

This follows from the definition of $I(C)$: the relation

$$I_t(C) = I_{t_{i-1}}(C) + Z_i\, (B_t - B_{t_{i-1}}), \quad t_{i-1} \le t \le t_i \,,$$

holds and the sample paths of Brownian motion are continuous.

2.2.3 The General Itô Stochastic Integral

In the previous section we introduced the Itô stochastic integral for simple processes C, i.e. for stochastic processes whose sample paths are step functions. The Itô stochastic integral $\int_0^t C_s\, dB_s$ is then simply the Riemann–Stieltjes sum of C, evaluated at the left end points of the intervals $[t_{i-1}, t_i]$, with respect to Brownian motion. We learnt in Section 2.1.2 that, in general, we cannot define the Itô stochastic integral as a pathwise limit of Riemann–Stieltjes sums. In Section 2.2.1 we suggested that we define the Itô stochastic integral as the mean square limit of suitable Riemann–Stieltjes sums. This idea works under quite general conditions, and it is the aim of this section to advocate this approach. At a certain point we will need some tools from Hilbert space theory, which we do not require as common knowledge. Then we will depend on some heuristic arguments. However, the interested reader can find a proof of the existence of the general Itô stochastic integral, using tools from measure theory and functional analysis, in Appendix A4.

In what follows, the process $C = (C_t, t \in [0, T])$ serves as the integrand of the Itô stochastic integral. We suppose that the following conditions are satisfied:

Assumptions on the Integrand Process C:

- C is adapted to Brownian motion on $[0, T]$, i.e. C_t is a function of B_s, $s \le t$.

- The integral $\int_0^T E C_s^2 \, ds$ is finite.

Notice that the Assumptions are trivially satisfied for a simple process; cf. p. 101. (Verify them!) Another class of admissible integrands consists of the deterministic functions $c(t)$ on $[0, T]$ with $\int_0^T c^2(t) \, dt < \infty$. It includes the continuous functions on $[0, T]$.

In what follows, we give a heuristic approach to the Itô stochastic integral. First recall how we proceeded for the definition of the integral $\int_0^t B_s \, dB_s$ in Section 2.2.1:

- For fixed t and a given partition (t_i) of $[0, t]$ we introduced the Riemann–Stieltjes sums $\sum_{i=1}^n B_{t_{i-1}} (B_{t_i} - B_{t_{i-1}})$.

- Notice: in Section 2.2.2 we defined exactly these Riemann–Stieltjes sums as the Itô stochastic integrals $I(C^{(n)}) = \int_0^t C_s^{(n)} \, dB_s$ of the simple functions $C^{(n)}$ which assume the value $B_{t_{i-1}}$ on the interval $[t_{i-1}, t_i)$.

- Then we considered the mean square limit of the Riemann–Stieltjes sums $I(C^{(n)})$ and obtained the value $0.5 \, (B_t^2 - t)$.

This approach can be made to work for processes C satisfying the Assumptions. We commence with an interesting observation:

Let C be a process satisfying the Assumptions. Then one can find a sequence $(C^{(n)})$ of simple processes such that

$$\int_0^T E[C_s - C_s^{(n)}]^2 \, ds \to 0 \,.$$

The proof of this property is beyond the scope of this book; see Appendix A4 for a reference.

Thus the simple processes $C^{(n)}$ converge in a certain mean square sense to the integrand process C. Since $C^{(n)}$ is simple we can evaluate the Itô stochastic integrals $I_t(C^{(n)}) = \int_0^t C_s^{(n)} \, dB_s$ for every n and t.

The next step is to show that the sequence $(I(C^{(n)}))$ of Itô stochastic integrals converges in a certain mean square sense to a unique limit process. Indeed, one can show the existence of a process $I(C)$ on $[0, T]$ such that

$$E \sup_{0 \le t \le T} \left[I_t(C) - I_t(C^{(n)}) \right]^2 \to 0 \,.$$

See Appendix A4 for a proof.

The mean square limit $I(C)$ is called the *Itô stochastic integral of C*. It is denoted by

$$I_t(C) = \int_0^t C_s \, dB_s \,, \quad t \in [0, T] \,.$$

For a simple process C, the Itô stochastic integral has the Riemann–Stieltjes sum representation (2.11).

After this general definition we do not feel very comfortable with the notion of Itô stochastic integral. We have lost the intuition because we are not able to write the integral $\int_0^t C_s \, dB_s$ in simple terms of Brownian motion. In particular cases we will be able to obtain explicit formulae for Itô stochastic integrals, but this requires knowledge of the Itô lemma; see Section 2.3. For our general intuition, and for practical purposes, the following rule of thumb is helpful:

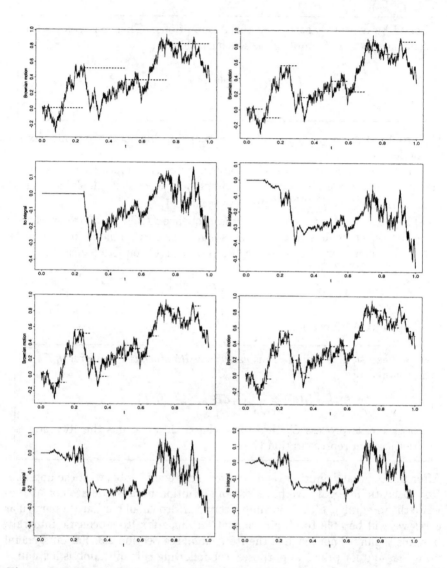

Figure 2.2.5 1st and 3rd rows: *approximations to a Brownian sample path by simple processes $C^{(n)}$ (dashed lines) assuming $n = 4, 10, 20, 40$ distinct values. Below every figure you find the path of the corresponding Itô stochastic integral $I(C^{(n)})$. The latter processes approximate the process $I_t(B) = \int_0^t B_s \, dB_s = 0.5 \, (B_t^2 - t)$. Compare with Figure 2.3.1, where the corresponding sample path of $I(B)$ is drawn.*

The Itô stochastic integrals $I_t(C) = \int_0^t C_s\, dB_s$, $t \in [0,T]$, constitute a stochastic process. For a given partition

$$\tau_n: \quad 0 = t_0 < t_1 < \cdots < t_{n-1} < t_n = T\,,$$

and $t \in [t_{k-1}, t_k]$, the random variable $I_t(C)$ is "close" to the Riemann–Stieltjes sum

$$\sum_{i=1}^{k-1} C_{t_{i-1}} \left(B_{t_i} - B_{t_{i-1}} \right) + C_{t_{k-1}} \left(B_t - B_{t_{k-1}} \right),$$

and this approximation is the closer (in the mean square sense) to the value of $I_t(C)$ the more dense the partition τ_n in $[0,T]$.

Properties of the General Itô Stochastic Integral

The general Itô stochastic integral inherits the properties of the Itô stochastic integral for simple processes, see Section 2.2.2. In general, we will not be able to prove these properties. However, some proofs can be found in Appendix A4.

The stochastic process $I_t(C) = \int_0^t C_s\, dB_s$, $t \in [0,T]$, is a martingale with respect to the natural Brownian filtration $(\mathcal{F}_t, t \in [0,T])$.

This results from the particular choice of the approximating Riemann–Stieltjes sums of C, which are evaluated at the left end points of the intervals $[t_{i-1}, t_i]$.

The Itô stochastic integral has expectation zero.

For the proof of the existence of the Itô stochastic integral the isometry property (2.14) for simple processes is essential. It remains valid in the general case.

The Itô stochastic integral satisfies the *isometry property* :

$$E\left(\int_0^t C_s\, dB_s \right)^2 = \int_0^t E C_s^2\, ds\,, \quad t \in [0,T]\,.$$

The Itô stochastic integral also has some properties in common with the Riemann and Riemann–Stieltjes integrals.

The Itô stochastic integral is *linear*:

For constants c_1, c_2 and processes $C^{(1)}$ and $C^{(2)}$ on $[0, T]$, satisfying the Assumptions,

$$\int_0^t \left[c_1\, C_s^{(1)} + c_2\, C_s^{(2)} \right] dB_s = c_1 \int_0^t C_s^{(1)}\, dB_s + c_2 \int_0^t C_s^{(2)}\, dB_s\,.$$

The Itô stochastic integral is *linear on adjacent intervals*:

For $0 \le t \le T$,

$$\int_0^T C_s\, dB_s = \int_0^t C_s\, dB_s + \int_t^T C_s\, dB_s\,.$$

Finally, we state the following property:

The process $I(C)$ has continuous sample paths.

Notes and Comments

The definition of the Itô stochastic integral goes back to Itô (1942,1944) who introduced the stochastic integral with a random integrand. Doob (1953) realized the connection of Itô integration and martingale theory. Itô integration is by now part of advanced textbooks on probability theory and stochastic processes. Standard references are Chung and Williams (1990), Ikeda and Watanabe (1989), Karatzas and Shreve (1988), Øksendahl (1985) and Protter (1992). Nowadays, some texts on finance also contain a survival kit on stochastic integration; see for example Lamberton and Lapeyre (1996).

Some of the aforementioned books define the stochastic integral with respect to processes more general than Brownian motion, including processes with jumps. Moreover, the assumption $\int_0^T EC_s^2\, ds < \infty$ for the existence of $\int_0^t C_s\, dB_s$ can be substantially relaxed.

2.3   The Itô Lemma

In the last two sections we learnt about the Itô stochastic integral. Now we know how the integral $\int_0^t C_s\, dB_s$ is defined, but with the exception of simple processes C we do not have the tools to calculate Itô stochastic integrals and to proceed with some simple operations on them. It is the objective of this section to provide such a tool, the Itô lemma.

2.3.1   The Classical Chain Rule of Differentiation

The Itô lemma is the stochastic analogue of the classical chain rule of differentiation. We mentioned the chain rule on p. 100 in a particular case. There we recalled that, for a differentiable function $b(s)$,

$$\frac{1}{2} \frac{db^2(s)}{ds} = b(s) \frac{db(s)}{ds} .$$

This relation implies that, with $b(0) = 0$,

$$\frac{1}{2} \int_0^t \frac{db^2(s)}{ds} \, ds = \frac{1}{2} b^2(t) = \int_0^t b(s) \frac{db(s)}{ds} \, ds = \int_0^t b(s) \, db(s) . \qquad (2.16)$$

We cannot simply replace in (2.16) the deterministic function $b(s)$ with a sample path $B_s(\omega)$ of Brownian motion. Indeed, such a path is nowhere differentiable. Using some elementary means, we discovered in Section 2.2.1 that the Itô stochastic integral $\int_0^t B_s \, dB_s$ has value $0.5 \, (B_t^2 - t)$. This is in contrast to (2.16). Notice that the integral value $0.5 \, (B_t^2 - t)$ is the value $0.5 \, B_t^2$, which is to be expected from the classical chain rule of differentiation, corrected by $-0.5 \, t$. This suggests that we have to find a correction term to the classical chain rule. Another appeal to the discussion in Section 2.2.1 tells us that this term comes from the mean square limit of the quadratic functional $\sum_{i=1}^n (B_{t_i} - B_{t_{i-1}})^2$.

In order to understand the *stochastic* chain rule, we first recall the *deterministic* chain rule. For simplicity, we write $h'(t)$, $h''(t)$, etc., for the ordinary derivatives of the function h at t.

Let f and g be differentiable functions. Then the *classical chain rule of differentiation* holds:

$$[f(g(s))]' = f'(g(s)) \, g'(s) . \qquad (2.17)$$

Because we are interested in integrals, we rewrite (2.17) in integral form, i.e. we integrate both sides of (2.17) on $[0, t]$.

The *chain rule in integral form* :

$$f(g(t)) - f(g(0)) = \int_0^t f'(g(s)) \, g'(s) \, ds = \int_0^t f'(g(s)) \, dg(s) . \qquad (2.18)$$

We give a heuristic argument for the validity of the chain rule, which we will take as a motivation for the Itô lemma. In the language of differentials, (2.17) becomes $df(g) = f' \, dg$. This differential can be interpreted as the first order term in a Taylor expansion:

$$f(g(t) + dg(t)) - f(g(t)) = f'(g(t)) \, dg(t) \; + \; \frac{1}{2} f''(g(t)) \, [dg(t)]^2 \; + \; \cdots . \quad (2.19)$$

Here $dg(t) = g(t + dt) - g(t)$ is the increment of g on $[t, t + dt]$. Under suitable conditions, the second and higher order terms in this expansion are negligible for small dt.

2.3.2 A Simple Version of the Itô Lemma

Now assume that f is a twice differentiable function, but replace $g(t)$ in (2.19) with a sample path $B_t(\omega)$ of Brownian motion. Write $dB_t = B_{t+dt} - B_t$ for the increment of B on $[t, t + dt]$. Using the same arguments as above, we obtain

$$f(B_t + dB_t) - f(B_t) = f'(B_t) \, dB_t \; + \; \frac{1}{2} f''(B_t) \, (dB_t)^2 \; + \; \cdots . \quad (2.20)$$

On p. 99 we gave some motivation that the squared differential $(dB_t)^2$ can be interpreted as dt. Thus,

in contrast to the deterministic case, the contribution of the second order term in the Taylor expansion (2.20) is not negligible.

This fact is *the* reason for the deviation from the classical chain rule.

Integrating both sides of (2.20) in a formal sense and neglecting terms of order higher than 3 on the right-hand side, we obtain for $s < t$,

$$\int_s^t df(B_x) \quad := \quad f(B_t) - f(B_s) \quad (2.21)$$

$$= \quad \int_s^t f'(B_x) \, dB_x \; + \frac{1}{2} \int_s^t f''(B_x) \, dx . \quad (2.22)$$

The following question naturally arises:

How do we have to interpret the integrals in (2.21) and (2.22)?

The first integral in (2.22) is the Itô stochastic integral of $f'(B)$, and the second integral has to be interpreted as the Riemann integral of $f''(B)$. Relation (2.21) defines the quantity $\int_s^t df(B_s)$ which is an analogue of a telescoping sum, and therefore it is natural to assign the value $f(B_t) - f(B_s)$ to it. In what follows, we always give the value $V_t - V_s$ to the symbolic integral $\int_s^t dV_x$, whatever the stochastic process V.

Let f be twice continuously differentiable.

The formula

$$f(B_t) - f(B_s) = \int_s^t f'(B_x)\, dB_x + \frac{1}{2} \int_s^t f''(B_x)\, dx, \quad s < t, \quad (2.23)$$

is a simple form of the *Itô lemma* or of the *Itô formula*.

Using sophisticated methods, which are beyond the scope of this book, one can show that (2.23) is mathematically correct.

Now we want to see the Itô lemma at work.

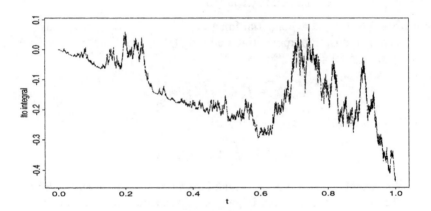

Figure 2.3.1 *One sample path of the process $\int_0^t B_s\, dB_s = 0.5\,(B_t^2 - t)$. See also Figure 2.2.5.*

Example 2.3.2 Choose $f(t) = t^2$. Then $f'(t) = 2t$ and $f''(t) = 2$. Hence the Itô lemma yields for $s < t$,

$$B_t^2 - B_s^2 = 2 \int_s^t B_x\, dB_x + \int_s^t dx.$$

For $s = 0$ we obtain the formula

$$\int_0^t B_x \, dB_x = \frac{1}{2} \left(B_t^2 - t \right),$$

which is already familiar to us; cf. (2.6) or the discussion in Section 2.3.1.

We try to get a similar formula for $B_t^3 - B_s^3$. In this case, $f(t) = t^3$, $f'(t) = 3t^2$ and $f''(t) = 6t$. The Itô lemma immediately yields

$$B_t^3 - B_s^3 = 3 \int_s^t B_x^2 \, dB_x + 3 \int_s^t B_x \, dx.$$

We cannot express $\int_s^t B_x \, dx$ in simpler terms of Brownian motion. In order to get an impression of the kind of sample paths of the processes $\int_s^t B_x \, dx$ or $\int_s^t B_x^2 \, dB_x$ one has to rely on simulations.

Now exercise the Itô lemma yourself: calculate $f(B_t) - f(B_s)$ for a power function $f(t) = t^n$ for some integer $n > 3$. $\qquad\qquad\qquad\qquad\qquad\qquad\square$

Example 2.3.3 (The exponential function is not the Itô exponential)
In classical calculus, the exponential function $f(t) = \exp\{t\}$ has the spectacular property that $f'(t) = f(t)$. Equivalently,

$$f(t) - f(s) = \int_s^t f(x) \, dx.$$

Is there a stochastic process X such that

$$X_t - X_s = \int_s^t X_x \, dB_x, \quad s < t, \tag{2.24}$$

in the Itô sense? This would be a candidate for the *Itô exponential*. Since the classical rules of integration fail in general when Brownian motion is involved, we may also expect that $f(B_t) = \exp\{B_t\}$ is not the Itô exponential. This can be checked quite easily: since $f(t) = f'(t) = f''(t)$ the Itô lemma yields for $s < t$,

$$e^{B_t} - e^{B_s} = \int_s^t e^{B_x} \, dB_x + \frac{1}{2} \int_s^t e^{B_x} \, dx.$$

Obviously, the second, Riemann, integral yields a positive value, and so (2.24) cannot be satisfied.

At the moment we cannot give an answer to the above question. We will return to this problem when we have a more advanced form of the Itô lemma; see Example 2.3.5. $\qquad\qquad\qquad\qquad\qquad\qquad\square$

2.3.3   Extended Versions of the Itô Lemma

In this section we extend the simple Itô lemma, given in (2.21)–(2.22), in various ways.

We start with an Itô lemma for the stochastic process $f(t, B_t)$. Assume that $f(t, x)$ has continuous partial derivatives of at least second order. A modification of the Taylor expansion argument used in Section 2.3.2 is also successful in this more general case.

Recall that a second order Taylor expansion yields that

$$f(t + dt, B_{t+dt}) - f(t, B_t) = \qquad\qquad\qquad\qquad (2.25)$$

$$f_1(t, B_t)\, dt \; + \; f_2(t, B_t)\, dB_t$$

$$+ \;\; \frac{1}{2} \left[f_{11}(t, B_t)\, (dt)^2 \; + \; 2\, f_{12}(t, B_t)\, dt\, dB_t \; + \; f_{22}(t, B_t)\, (dB_t)^2 \right]$$

$$+ \;\;\; \cdots \;\; .$$

Here, and in what follows, we use the following notation for the partial derivatives of f:

$$f_i(t, x) = \left. \frac{\partial}{\partial x_i}\, f(x_1, x_2) \right|_{x_1 = t, x_2 = x} , \qquad i = 1, 2,$$

$$f_{ij}(t, x) = \left. \frac{\partial}{\partial x_i} \frac{\partial}{\partial x_j} f(x_1, x_2) \right|_{x_1 = t, x_2 = x} , \qquad i, j = 1, 2 .$$

As in classical calculus, higher order terms in (2.25) are negligible, and so are the terms with factors $dt\, dB_t$ and $(dt)^2$. However, since we interpret $(dB_t)^2$ as dt, the term with $(dB_t)^2$ cannot be neglected. Arguing in the same way as in Section 2.3.2, i.e. formally integrating the left-hand and right-hand sides in (2.25) and collecting all terms with dt and dB_t separately, we end up with the following formula:

Extension I of the Itô Lemma:

Let $f(t, x)$ be a function whose second order partial derivatives are continuous. Then

$$f(t, B_t) - f(s, B_s) = \qquad\qquad\qquad\qquad (2.26)$$

$$\int_s^t \left[f_1(x, B_x) + \frac{1}{2} f_{22}(x, B_x) \right] dx \; + \; \int_s^t f_2(x, B_x)\, dB_x , \quad s < t .$$

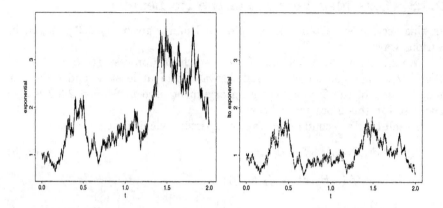

Figure 2.3.4 *One sample path of the exponential of Brownian motion* $\exp\{B_t\}$ (left) *and of the Itô exponential* $\exp\{B_t - 0.5\,t\}$ *(right) for the same path of B; cf. Examples 2.3.3 and 2.3.5.*

We apply this formula to find the Itô exponential:

Example 2.3.5 (The Itô exponential)
We learnt in Example 2.3.3 that the stochastic process $\exp\{B_t\}$ is not the Itô exponential in the sense of (2.24). Now we choose the function

$$f(t, x) = e^{x - 0.5\,t}\,.$$

Then direct calculation shows that

$$f_1(t, x) = -\frac{1}{2} f(t, x)\,, \quad f_2(t, x) = f(t, x)\,, \quad f_{22}(t, x) = f(t, x)\,.$$

An application of the above Itô lemma gives

$$f(t, B_t) - f(s, B_s) = \int_s^t f(x, B_x)\, dB_x\,.$$

In the sense of (2.24), $f(t, B_t)$ is the *Itô exponential*. See Figure 2.3.4 for a comparison of the paths of the processes $\exp\{B_t\}$ and $\exp\{B_t - 0.5\,t\}$.   □

Example 2.3.6 (Geometric Brownian motion)
Consider a particular form of geometric Brownian motion (cf. Example 1.3.8):

$$X_t = f(t, B_t) = e^{(c - 0.5\,\sigma^2)\,t + \sigma\,B_t}\,, \tag{2.27}$$

where c and $\sigma > 0$ are constants. Notice that

$$f(t, x) \;=\; \mathrm{e}^{(c-0.5\,\sigma^2)\,t+\sigma x}, \quad f_1(t, x) \;=\; (c - 0.5\,\sigma^2)\,f(t, x),$$

$$f_2(t, x) \;=\; \sigma\,f(t, x), \qquad f_{22}(t, x) \;=\; \sigma^2\,f(t, x).$$

An application of the Itô lemma (2.26) yields that the process X satisfies the linear stochastic differential equation

$$X_t - X_0 = c \int_0^t X_s\,ds \;+\; \sigma \int_0^t X_s\,dB_s. \tag{2.28}$$

$\square$

For later use we will need an even more general version of the Itô lemma. We will consider processes of the form $f(t, X_t)$, where X is given by

$$X_t = X_0 + \int_0^t A_s^{(1)}\,ds \;+\; \int_0^t A_s^{(2)}\,dB_s, \tag{2.29}$$

and both, $A^{(1)}$ and $A^{(2)}$, are adapted to Brownian motion. Here it is assumed that the above integrals are well defined in the Riemann and Itô senses, respectively.

> A process X, which has representation (2.29), is called an *Itô process*. One can show that the processes $A^{(1)}$ and $A^{(2)}$ are uniquely determined in the sense that, if X has a representation (2.29), where the $A^{(i)}$s are replaced with adapted processes $D^{(i)}$, then $A^{(i)}$ and $D^{(i)}$ necessarily coincide.

Recall that the geometric Brownian motion (2.27) satisfies (2.28). Hence it is an Itô process with $A^{(1)} = cX$ and $A^{(2)} = \sigma X$.

Now, using a similar argument with a Taylor expansion as above, one can show the following formula.

Extension II of the Itô Lemma:

Let X be an Itô process with representation (2.29) and $f(t, x)$ be a function whose second order partial derivatives are continuous. Then

$$f(t, X_t) - f(s, X_s) = \qquad\qquad\qquad\qquad\qquad (2.30)$$

$$\int_s^t \left[f_1(y, X_y) + A_y^{(1)} f_2(y, X_y) + \frac{1}{2} [A_y^{(2)}]^2 f_{22}(y, X_y) \right] dy$$

$$+ \int_s^t A_y^{(2)} f_2(y, X_y) \, dB_y, \quad s < t.$$

Give a justification of this formula. Use a Taylor expansion for $f(t+dt, X_{t+dt}) - f(t, X_t)$ as in (2.25), where B is replaced with X, and X has representation (2.29). Neglect higher order terms, starting with terms involving $(dt)^2$ and $dt\, dB_t$, and make use of $(dB_t)^2 = dt$.

Formula (2.30) is frequently given in the following form:

$$f(t, X_t) - f(s, X_s) \qquad\qquad\qquad\qquad\qquad (2.31)$$

$$= \int_s^t \left[f_1(y, X_y) + \frac{1}{2} [A_y^{(2)}]^2 f_{22}(y, X_y) \right] dy + \int_s^t f_2(y, X_y) \, dX_y,$$

where

$$dX_y = A_y^{(1)} \, dy + A_y^{(2)} \, dB_y.$$

The latter identity is a symbolic way of writing the Itô representation (2.29). The integral with respect to X in (2.31) has to be interpreted as follows:

$$\int_s^t f_2(y, X_y) \, dX_y = \int_s^t A_y^{(1)} f_2(y, X_y) \, dy + \int_s^t A_y^{(2)} f_2(y, X_y) \, dB_y. \quad (2.32)$$

Finally, we consider the Itô lemma for stochastic processes of the form $f(t, X_t^{(1)}, X_t^{(2)})$, where both, $X^{(1)}$ and $X^{(2)}$, are Itô processes with respect to the same Brownian motion:

$$X_t^{(i)} = X_0^{(i)} + \int_0^t A_s^{(1,i)} \, ds + \int_0^t A_s^{(2,i)} \, dB_s, \quad i = 1, 2. \qquad (2.33)$$

A Taylor series expansion argument similar to the one above yields the following formula:

Extension III of the Itô Lemma:

Let $X^{(1)}$ and $X^{(2)}$ be two Itô processes given by (2.33) and $f(t, x_1, x_2)$ be a function whose second order partial derivatives are continuous. Then for $s < t$,

$$f(t, X_t^{(1)}, X_t^{(2)}) - f(s, X_s^{(1)}, X_s^{(2)}) \tag{2.34}$$

$$= \int_s^t f_1(y, X_y^{(1)}, X_y^{(2)}) \, dy$$

$$+ \sum_{i=2}^{3} \int_s^t f_i(y, X_y^{(1)}, X_y^{(2)}) \, dX_y^{(i)}$$

$$+ \frac{1}{2} \sum_{i=2}^{3} \sum_{j=2}^{3} \int_s^t f_{ij}(y, X_y^{(1)}, X_y^{(2)}) \, A_y^{(2,i)} A_y^{(2,j)} \, dy \, .$$

Here $f_i(t, x_1, x_2)$, $f_{ij}(t, x_1, x_2)$ are the partial derivatives of $f(t, x_1, x_2)$ with respect to the ith, the ith and jth variables, respectively; cf. p. 117.

The integrals with respect to $X_y^{(i)}$ have to be interpreted in the same way as in (2.32). Formula (2.34) can be extended in the straightforward way to functions $f(t, X_t^{(1)}, \ldots, X_t^{(m)})$, where the $X^{(i)}$s are Itô processes with respect to the same Brownian motion. We mention that one can also consider such a formula for Itô processes with respect to different Brownian motions. However, this requires that we define the Itô stochastic integral for such a setting.

The following example is an application of the Itô lemma (2.34).

Example 2.3.7 (Integration by parts)
Consider the function $f(t, x_1, x_2) = x_1 x_2$. Then we obtain (we suppress the arguments of the functions)

$$f_1 = 0, \quad f_2 = x_2, \quad f_3 = x_1, \quad f_{22} = f_{33} = 0 \quad \text{and} \quad f_{23} = f_{32} = 1 \, .$$

Now apply formula (2.34) to obtain:

> **The Integration by Parts Formula:**
>
> Let $X^{(1)}$ and $X^{(2)}$ be two Itô processes given by (2.33). Then
>
> $$d(X_t^{(1)} X_t^{(2)}) = X_t^{(2)} \, dX_t^{(1)} + X_t^{(1)} \, dX_t^{(2)} + A_t^{(2,1)} A_t^{(2,2)} \, dt \,. \qquad (2.35)$$

Using the Itô representation (2.33), we obtain an alternative expression for the differentials:

$$
\begin{aligned}
d(X_t^{(1)} X_t^{(2)}) = {} & (X_t^{(1)} A_t^{(1,2)} + X_t^{(2)} A_t^{(1,1)} + A_t^{(2,1)} A_t^{(2,2)}) \, dt \\
& + (X_t^{(1)} A_t^{(2,2)} + X_t^{(2)} A_t^{(2,1)}) \, dB_t \,.
\end{aligned}
$$

As particular examples we consider

$$X_t^{(1)} = e^t - 1 = \int_0^t e^s \, ds \quad \text{and} \quad X_t^{(2)} = B_t = \int_0^t 1 \, dB_s \,.$$

Obviously,

$$A_t^{(1,1)} = e^t \,, \quad A_t^{(2,1)} = 0 \,, \quad A_t^{(1,2)} = 0 \,, \quad A_t^{(2,2)} = 1 \,.$$

Hence, integration by parts yields

$$\int_0^t e^s \, dB_s = e^t B_t - \int_0^t B_s \, e^s \, ds \,.$$

More generally, show that for any continuously differentiable function f on $[0, T]$ the following relation holds:

$$\int_0^t f(s) \, dB_s = f(t) B_t - \int_0^t f'(s) \, B_s \, ds \,. \qquad \qquad \square$$

Notes and Comments

The Itô lemma is the most important tool in Itô calculus. We will use it very often in the following sections. A first version of this fundamental result was proved by Itô (1951). Various versions of the Itô lemma and their proofs can be found in textbooks on stochastic calculus, for example, in Chung and Williams (1990), Ikeda and Watanabe (1989), Karatzas and Shreve (1988), Øksendahl (1985) or Protter (1992).

2.4 The Stratonovich and Other Integrals

In this section we discuss some other stochastic integrals and their relation with the Itô stochastic integral. It is certainly useful for you to realize that there is a large variety of other integrals, the Itô integral being just one member of this family. On the other hand, you do not need the information of this section (which is also slightly technical) unless you are interested in solving Itô stochastic differential equations by the so-called Stratonovich calculus.

In the previous sections we studied the Itô stochastic integrals

$$I_t(C) = \int_0^t C_s \, dB_s \,, \quad t \in [0, T] \,.$$

Here and in what follows, $B = (B_t, t \geq 0)$ is Brownian motion and $C = (C_t, t \in [0, T])$ is a process adapted to the natural Brownian filtration $\mathcal{F}_t = \sigma(B_s, s \leq t)$, $t \in [0, T]$. The main point in the definition of the Itô stochastic integral was the approximation of $I_t(C)$ by Riemann–Stieltjes sums of the form

$$\sum_{i=1}^{k-1} C_{t_{i-1}} \Delta_i B + C_{t_{k-1}} (B_t - B_{t_{k-1}}) \quad \text{for} \quad t_{k-1} \leq t \leq t_k \,, \tag{2.36}$$

for partitions

$$\tau_n : \quad 0 = t_0 < t_1 < \cdots < t_{n-1} < t_n = T$$

such that

$$\text{mesh}(\tau_n) = \max_{i=1,\ldots,n} (t_i - t_{i-1}) \to 0 \,.$$

As usual, we write $\Delta_i B = B_{t_i} - B_{t_{i-1}}$.

In the Riemann–Stieltjes sums (2.36) we chose the values of C at the left end points of the subintervals $[t_{i-1}, t_i]$. This choice was made for mathematical convenience. Our gain was that the stochastic process $(I_t(C), t \in [0, T])$ has a rich mathematical structure. It inherits the martingale property from the approximating Riemann–Stieltjes sums (2.36). Thus a high level of the theory of stochastic processes can be applied. As subsidiary properties we get that the expectation of the Itô stochastic integral is always zero and its variance can be expressed by an application of the isometry property. The price one has to pay for the nice mathematical structure is that the chain rule of classical calculus is not valid anymore. In Itô calculus, the Itô lemma replaces the classical chain rule.

In this section we consider another type of stochastic integral, the *Stratonovich stochastic integral*. We start with its definition in the case that the integrand process C is given by

$$C_t = f(B_t), \quad t \in [0, T],$$

for a twice differentiable function f on $[0, T]$. Define the Riemann–Stieltjes sums

$$\widetilde{S}_n = \sum_{i=1}^{n} f(B_{y_i}) \Delta_i B, \tag{2.37}$$

where

$$y_i = \frac{t_{i-1} + t_i}{2}, \quad i = 1, \dots, n.$$

One can show that the mean square limit of the Riemann-Stieltjes sums $\widetilde{S}_n$ exists if $\operatorname{mesh}(\tau_n) \to 0$.

The unique mean square limit $S_T(f(B))$ of the Riemann–Stieltjes sums $\widetilde{S}_n$ exists if $\int_0^T E f^2(B_t)\, dt < \infty$. The limit is called the *Stratonovich stochastic integral of $f(B)$*. It is denoted by

$$S_T(f(B)) = \int_0^T f(B_s) \circ dB_s.$$

Clearly, as for the Itô stochastic integral, we can define the stochastic process of the Stratonovich stochastic integrals

$$S_t(f(B)) = \int_0^t f(B_s) \circ dB_s, \quad t \in [0, T],$$

as the mean square limit of the corresponding Riemann–Stieltjes sums.

Example 2.4.1 On p. 100 we considered the Riemann–Stieltjes sums

$$\widetilde{S}_n = \sum_{i=1}^{n} B_{y_i} \Delta_i B$$

for a given partition τ_n of $[0, T]$. We also mentioned that it is not difficult to verify that the sequence $(\widetilde{S}_n)$ has the mean square limit $0.5 B_T^2$. This is the value of the corresponding Stratonovich stochastic integral:

$$S_T(B) = \int_0^T B_s \circ dB_s = \frac{1}{2} B_T^2. \tag{2.38}$$

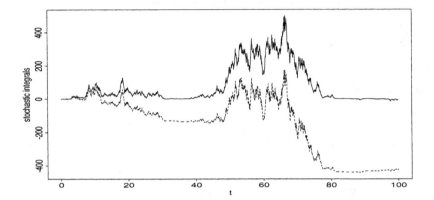

Figure 2.4.2 *One sample path of the process* $\int_0^t B_s^2 \, dB_s$ *(lower curve) and of the process* $\int_0^t B_s^2 \circ dB_s$ *(upper curve) for the same Brownian sample path.*

Obviously, the stochastic process $(0.5\, B_t^2, t \in [0, T])$ is not a martingale. However, we see that the Stratonovich stochastic integral (2.38) *formally* satisfies the classical chain rule; see p. 100 for a discussion. □

The formal validity of the classical chain rule is *the* reason for the use of Stratonovich stochastic integrals. Thus, despite the "poor" mathematical structure of the Stratonovich stochastic integral, it has this "nice" property. We will exploit it when we solve Itô stochastic differential equations.

To get some feeling for the Stratonovich stochastic integral, we consider a transformation formula which links the Itô and the corresponding Stratonovich stochastic integrals with the same integrand process $f(B)$. We assume

$$\int_0^T E[f(B_t)]^2 \, dt < \infty \quad \text{and} \quad \int_0^T E[f'(B_t)]^2 \, dt < \infty. \tag{2.39}$$

The first condition is needed for the definition of the Stratonovich stochastic integral $S_T(f(B))$ and the Itô stochastic integral $I_T(f(B))$.

First observe that the Taylor expansion

$$f(B_{y_i}) = f(B_{t_{i-1}}) + f'(B_{t_{i-1}}) \left(B_{y_i} - B_{t_{i-1}} \right) + \cdots$$

holds, where we neglect higher order terms. Thus an appoximating Riemann–

Stieltjes sum for $S_T(f(B))$ can be written as follows:

$$\sum_{i=1}^{n} f(B_{y_i}) \Delta_i B$$

$$= \sum_{i=1}^{n} f(B_{t_{i-1}}) \Delta_i B + \sum_{i=1}^{n} f'(B_{t_{i-1}}) (B_{y_i} - B_{t_{i-1}}) \Delta_i B + \cdots$$

$$= \sum_{i=1}^{n} f(B_{t_{i-1}}) \Delta_i B + \sum_{i=1}^{n} f'(B_{t_{i-1}}) (B_{y_i} - B_{t_{i-1}})^2$$

$$+ \sum_{i=1}^{n} f'(B_{t_{i-1}})(B_{y_i} - B_{t_{i-1}})(B_{t_i} - B_{y_i}) + \cdots$$

$$= \widetilde{S}_n^{(1)} + \widetilde{S}_n^{(2)} + \widetilde{S}_n^{(3)} + \cdots, \tag{2.40}$$

where we again neglected higher order terms. By definition of the Itô integral, $\widetilde{S}_n^{(1)}$ has the mean square limit $\int_0^T f(B_s)\, dB_s$. An application of condition (2.39) shows that $\widetilde{S}_n^{(3)}$ has the mean square limit zero, and some calculations give that

$$\widetilde{S}_n^{(2)} \to \frac{1}{2} \int_0^T f'(B_t)\, dt$$

in mean square. Combining the mean square convergences of $S_n^{(i)}$, $i = 1, 2, 3$, we finally obtain the following result from (2.40):

Assume the function f satisfies (2.39). Then the *transformation formula* holds:

$$\int_0^T f(B_t) \circ dB_t = \int_0^T f(B_t)\, dB_t + \frac{1}{2} \int_0^T f'(B_t)\, dt. \tag{2.41}$$

From this formula it is immediate that $(S_t(f(B)), t \in [0, T])$ does not constitute a martingale. Check this by taking expectations in (2.41).

Now take the particular function $f(t) = g'(t)$. An application of the Itô lemma (2.23) to $Y_t = g(B_t)$ yields

$$g(B_T) - g(B_0) = \int_0^T g'(B_s)\, dB_s + \frac{1}{2} \int_0^T g''(B_s)\, ds$$

$$= \int_0^T f(B_s)\, dB_s + \frac{1}{2} \int_0^T f'(B_s)\, ds,$$

and the right-hand side is equal to $\int_0^T g'(B_s) \circ dB_s$. And so we have:

The Stratonovich stochastic integral satisfies the chain rule of classical calculus:

$$\int_0^T g'(B_s) \circ dB_s = g(B_T) - g(B_0). \qquad (2.42)$$

Notice: this statement does *not* mean that the Stratonovich stochastic integral is a classical (i.e. Riemann) integral. We only claim that the corresponding chain rules have a similar structure.

Example 2.4.3 a) Take $g(t) = t^2$. Then $g'(t) = 2t$. We obtain from (2.42) that

$$\int_0^T B_t \circ dB_t = \frac{1}{2} B_T^2 - \frac{1}{2} B_0^2 = \frac{1}{2} B_T^2.$$

This is in agreement with our previous observations about the value of $S_T(B)$.
b) Take $g(t) = \exp\{t\}$. Then $g'(t) = g(t)$. We obtain from (2.42) that

$$\int_0^T e^{B_t} \circ dB_t = e^{B_T} - e^{B_0} = e^{B_T} - 1.$$

We conclude from the latter relation that the process $X_t = \exp\{B_t\}$, $t \in [0, T]$, is the *Stratonovich exponential*. Also recall from Example 2.3.5 that the Itô exponential is a totally different process. $\qquad \square$

In what follows, we want to give a transformation formula for the Stratonovich stochastic integral which is more general than (2.41). There we only addressed integrand processes $C = f(B)$. Now we assume that the integrand is of the form

$$C_t = f(t, X_t), \quad t \in [0, T], \qquad (2.43)$$

where $f(t, x)$ is a function with continuous partial derivatives of order two. The process X is supposed to be an Itô process (see p. 119) given by the stochastic differential equation:

$$X_t = X_0 + \int_0^t a(s, X_s)\, ds + \int_0^t b(s, X_s)\, dB_s,$$

where the continuous functions $a(t, x)$ and $b(t, x)$ satisfy the existence and uniqueness conditions on p. 138. For integrands (2.43) it is not straightforward how to get an extension of the integral with $C = f(B)$. One possible way to define the Stratonovich stochastic integral is via approximating Riemann–Stieltjes sums

$$\widetilde{S}_n = \sum_{i=1}^{n} f(t_{i-1}, 0.5\,(X_{t_{i-1}} + X_{t_i}))\,\Delta_i B\,.$$

The mean square limit $\int_0^T f(t, X_t) \circ dB_t$ of these Riemann–Stieltjes sums exists if

$$\int_0^T E[f(t, X_t)]^2\, dt < \infty\,.$$

One can show that this definition is consistent with the previous one (with $f(t, x) = f(x)$, $X = B$ and the approximating Riemann–Stieltjes sums (2.37)).

For later use, we give here another identity:

Under the assumptions on f and X given above, the following *transformation formula* holds:

$$\int_0^T f(t, X_t) \circ dB_t \qquad\qquad\qquad\qquad (2.44)$$

$$= \int_0^T f(t, X_t)\, dB_t + \frac{1}{2} \int_0^T b(t, X_t)\, f_2(t, X_t)\, dt\,,$$

where, as usual, $f_2(t, x)$ is the partial derivative of f with respect to x.

From the above discussion it might have become clear that we can define very different stochastic integrals. For every $p \in [0, 1]$, a given partition $\tau_n = (t_i)$ of $[0, T]$ and a process C adapted to Brownian motion, we can define the Riemann–Stieltjes sums

$$\widetilde{S}_n^{(p)} = \sum_{i=1}^{n} C_{y_i(p)}\, \Delta_i B\,,$$

where

$$y_i(p) = t_{i-1} + p\,(t_i - t_{i-1})\,, \quad i = 1, \ldots, n\,.$$

If mesh$(\tau_n) \to 0$, the mean square limit of the Riemann–Stieltjes sums $\widetilde{S}_n^{(p)}$ exists, provided C satisfies some further assumptions. The limit can be considered as the (p)-*stochastic integral* of C:

$$I_T^{(p)}(C) = (p) - \int_0^T C_s \, dB_s .$$

We studied two particular cases: $p = 0$ (the Itô case) and $p = 0.5$ (the Stratonovich case). For non-trivial integrands C the values $I_T^{(p)}(C)$ differ for distinct ps. For example, using arguments similar to the Itô and Stratonovich cases, one can show that

$$(p) - \int_0^T B_s \, dB_s = \frac{1}{2} B_T^2 + \left(p - \frac{1}{2} \right) T .$$

It is also possible to get a transformation formula such as (2.44) in order to relate the (p)-integrals, $p \in [0, 1]$, to the corresponding Itô or Stratonovich integrals. For applications, the Itô and the Stratonovich integrals are most relevant. The reasons were explained above. See also Chapter 3 on stochastic differential equations.

Notes and Comments

The Stratonovich stochastic integral was introduced by Fisk, and independently by Stratonovich (1966). The mathematical theory of the Stratonovich integral can be found in Stratonovich (1966); see also Arnold (1973).

Both, the Itô and Stratonovich integrals, are defined in a mathematically correct way. In applications one has to make a decision about which stochastic integral is appropriate. This is then a question of modelling; see also the discussion on p. 150.

In Section 3.2.3 we will use the rules of Stratonovich calculus (i.e. the rules of classical calculus) for the solution of Itô stochastic differential equations.

3

Stochastic Differential Equations

This chapter is devoted to stochastic differential equations and their solution. In Section 3.1 we start with a short introduction to ordinary differential equations. Stochastic differential equations can be understood as deterministic differential equations which are perturbed by random noise.

In Section 3.2.1 we introduce Itô stochastic differential equations and explain what a solution is. It will turn out that stochastic differential equations are actually stochastic integral equations which involve ordinary and Itô stochastic integrals. Therefore the notion of "differential equation" might be slightly misleading, but since it is the general custom to use this term, we will follow it. In Section 3.2.1 we formulate conditions for the existence and uniqueness of solutions to Itô stochastic differential equations. In Section 3.2.2 we give a simple method for solving Itô stochastic differential equations. It is based on the Itô lemma. We continue in Section 3.2.3 with the solution of Itô stochastic differential equations which are derived from an equivalent Stratonovich stochastic differential equation.

In Section 3.3 we consider the general linear stochastic differential equation. It is a special class of stochastic differential equations which have an explicit solution in terms of the coefficient functions and of the underlying Brownian motion. Linear stochastic differential equations are relevant in many applications. We also provide a method for calculating the expectation and variance functions in this case.

As for deterministic differential equations, it is very exceptional to obtain an explicit solution to a stochastic differential equation. In general, one has to rely on numerical methods; see Section 3.4 for an introduction to this topic.

3.1   Deterministic Differential Equations

The theory of differential equations is the cradle of classical calculus. Such equations were the motivating examples for the creation of the differential and integral calculus. The idea underlying a differential equation is simple: we are given a functional relationship

$$f(t, x(t), x'(t), x''(t), \ldots) = 0, \quad 0 \le t \le T, \tag{3.1}$$

involving the time t, an *unknown* function $x(t)$ and its derivatives. It is the aim to find a function $x(t)$ which satisfies (3.1). It is called a *solution of the differential equation* (3.1). Hopefully, this solution is *unique* for a given *initial condition* $x(0) = x_0$, say.

The simplest differential equations are those of order 1. They involve only t, $x(t)$ and the first derivative $x'(t)$. Ideally, such a differential equation is given in the form

$$x'(t) = \frac{dx(t)}{dt} = a(t, x(t)), \quad x(0) = x_0, \tag{3.2}$$

for a known function $a(t, x)$. It is standard to write (3.2) in the more intuitive form

$$dx(t) = a(t, x(t)) \, dt, \quad x(0) = x_0. \tag{3.3}$$

If you interpret $x(t)$ as the location of a one-dimensional particle in space at time t, (3.3) describes the change of location of the particle in a small time interval $[t, t+dt]$, say. Relation (3.3) then tells us that $dx(t) = x(t+dt) - x(t)$ is proportional to the time increment dt with factor $a(t, x(t))$. Alternatively, (3.2) tells us that the velocity $x'(t)$ of the particle is a given function of time t and location $x(t)$.

Example 3.1.1 (Some simple differential equations)
Assume that the velocity is only a function of t:

$$x'(t) = a(t).$$

Integration on both sides yields the solution

$$x(t) = x(0) + \int_0^t a(s) \, ds.$$

This is the simplest form of a differential equation, but it is also a trivial one because the right-hand side does not depend on the unknown function $x(t)$.

Now assume that the velocity $x'(t)$ is proportional to the location $x(t)$:

$$x'(t) = c\,x(t)$$

for some constant c. Here simple integration does not help. But we know the solution to this differential equation: it is the exponential function $x(t) = x(0)\exp\{c\,t\}$. □

With an elementary trick one can sometimes solve the differential equation (3.2).

Example 3.1.2 (Separation of variables)
Suppose that the right-hand side of (3.2) can be separated into a product of two functions:

$$x'(t) = a_1(t)\,a_2(x(t))\,.$$

Symbolically, we rewrite this differential equation as

$$\frac{dx}{a_2(x)} = a_1(t)\,dt\,. \tag{3.4}$$

Now integrate both sides:

$$\int_{x(0)}^{x(t)} \frac{dx}{a_2(x)} = \int_0^t a_1(s)\,ds\,. \tag{3.5}$$

On the left-hand side we obtain a function of $x(t)$, on the right-hand side a function of t. Hopefully, we obtain an explicit form of the function $x(t)$.

This approach is justified by differentiating both sides of (3.5) as integrals of the upper limits. Then we obtain

$$\frac{x'(t)}{a_2(x(t))} = a_1(t)\,,$$

which is another form of writing (3.4).

We apply this method to the differential equation $x'(t) = c\,x(t)$ for some $x(0) \neq 0$. Then

$$\int_{x(0)}^{x(t)} \frac{dx}{x} = c\int_0^t ds\,,$$

giving $x(t) = x(0)\exp\{ct\}$. This is the solution we guessed in Example 3.1.1. □

From the above discussion we learnt about some of the important aspects of ordinary differential equations:

- *Solutions of differential equations are functions.* They describe the evolution or dynamics of a real-life process over a given period of time.

- In order to obtain a *unique* solution, one has to know the *initial condition* $x(0) = x_0$. If this unique solution $x(t)$ starts from a point x_0 at the present $t = 0$, say, the function $x(t)$ is completely determined in the future, i.e. for $t > 0$.

- Explicit solutions of differential equations are the exception from the rule. In general, one has to rely on numerical solutions to differential equations.

- Integrating both sides of the differential equation (3.2), one obtains an equivalent *integral equation*:

$$x(t) = x(0) + \int_0^t a\big(s, x(s)\big)\, ds \,.$$

Although this transformed equation is in general not of great use for finding the solution of (3.4), it gives an idea of how we could define a stochastic differential equation: as a stochastic integral equation.

3.2 Itô Stochastic Differential Equations

3.2.1 What is a Stochastic Differential Equation?

Consider the deterministic differential equation

$$dx(t) = a\big(t, x(t)\big)\, dt \,, \quad x(0) = x_0 \,.$$

The easiest way to introduce randomness in this equation is to randomize the initial condition. The solution $x(t)$ then becomes a stochastic process $(X_t, t \in [0, T])$:

$$dX_t = a(t, X_t)\, dt \,, \quad X_0(\omega) = Y(\omega) \,.$$

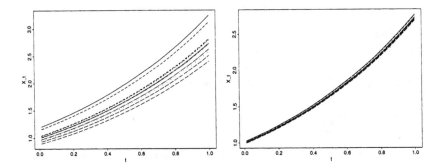

Figure 3.2.1 10 *solutions* $X_t = X_0 e^t$ *of the random differential equation* $dX_t = X_t\,dt$ *with initial condition* $X_0 = \exp\{N\}$, *where* N *has an* $N(0,\sigma^2)$ *distribution.* Left: $\sigma^2 = 0.01$. Right: $\sigma^2 = 0.0001$.

Such an equation is called a *random differential equation.* Its solution does not require stochastic calculus; we can use the classical methods and adjust the solution to the corresponding outcome of the initial condition. Random differential equations can be considered as deterministic differential equations with a perturbed initial condition. Their investigation can be of interest if one wants to study the robustness of the solution to a differential equation under a small change of the initial condition. For example, Figure 3.2.1 shows that the solution of a differential equation can change quite drastically, even if the change of the initial condition is small.

For our purposes, the randomness in the differential equation is introduced via an additional random noise term:

$$dX_t = a(t, X_t)\,dt + b(t, X_t)\,dB_t, \quad X_0(\omega) = Y(\omega). \tag{3.6}$$

Here, as usual, $B = (B_t, t \geq 0)$ denotes Brownian motion, and $a(t, x)$ and $b(t, x)$ are deterministic functions. The solution X, if it exists, is then a stochastic process. The randomness of $X = (X_t, t \in [0, T])$ results, on the one hand, from the initial condition, and on the other hand, from the noise generated by Brownian motion.

A naive interpretation of (3.6) tells us that the change $dX_t = X_{t+dt} - X_t$ is caused by a change dt of time, with factor $a(t, X_t)$, in combination with a change $dB_t = B_{t+dt} - B_t$ of Brownian motion, with factor $b(t, X_t)$. Since Brownian motion does not have differentiable sample paths (see Section 1.3.1), the following question naturally arises:

In which sense can we interpret (3.6)?

Clearly, there is no unique answer to this question. But since we know about Itô calculus, we can propose the following:

Interpret (3.6) as the stochastic integral equation

$$X_t = X_0 + \int_0^t a(s, X_s)\, ds + \int_0^t b(s, X_s)\, dB_s\,, \quad 0 \leq t \leq T\,, \qquad (3.7)$$

where the first integral on the right-hand side is a Riemann integral, and the second one is an Itô stochastic integral.

Equation (3.7) is called an *Itô stochastic differential equation.*

To call equation (3.7) a *differential* equation is counterintuitive. Its name originates from the symbolic equation (3.6), but the latter is meaningless unless one says what the differentials are. We follow the general custom and call (3.7) an Itô stochastic differential equation.

Brownian motion B is called the *driving process* of the Itô stochastic differential equation (3.7).

It is possible to replace Brownian motion by other driving processes, but this requires to define more general stochastic integrals. This is not a topic of this book. We refer to the monographs by Chung and Williams (1990) or Protter (1992) who introduce the stochastic integral with respect to semimartingales. The latter class of processes contains Brownian motion, but also a large variety of jump processes. They are useful tools when one is interested in modelling the jump character of real-life processes, for example, the strong oscillations of foreign exchange rates or crashes of the stock market.

It is a priori not clear whether the integrals in (3.7) are well defined. For example, can we ensure that $b(s, X_s)$ is adapted to Brownian motion and is $\int_0^T E[b(s, X_s)]^2\, ds$ finite? These are the crucial conditions for the definition of an Itô stochastic integral; see p. 108. But can one check these conditions if we do not know the solution X? Soon we will see that there exist simple conditions for the existence and uniqueness of solutions to Itô stochastic differential equations.

We first attempt the question:

What is a solution *of the Itô stochastic differential equation (3.7)?*

Surprisingly, there is no unique answer. We will find two kinds of solutions to a stochastic differential equation. These are called strong and weak solutions.

A *strong solution to the Itô stochastic differential equation* (3.7) is a stochastic process $X = (X_t, t \in [0, T])$ which satisfies the following conditions:

- X is adapted to Brownian motion, i.e. at time t it is a function of B_s, $s \le t$.

- The integrals occurring in (3.7) are well defined as Riemann or Itô stochastic integrals, respectively.

- X is a function of the underlying Brownian sample path and of the coefficient functions $a(t, x)$ and $b(t, x)$.

Thus a *strong* solution to (3.7) is based on the path of the underlying Brownian motion. If we were to change the Brownian motion by another Brownian motion we would get another strong solution which would be given by the same functional relationship, but with the new Brownian motion in it.

But what is now a weak *solution?*

For these solutions the path behavior is not essential, we are only interested in the distribution of X. The initial condition X_0 and the coefficient functions $a(t, x)$ and $b(t, x)$ are given, and we have to find a Brownian motion such that (3.7) holds. We mention that there exist Itô stochastic differential equations which have *only* weak solutions; see Chung and Williams (1990), p. 248, for an example.

Weak solutions X are sufficient in order to determine the distributional characteristics of X, such as the expectation, variance and covariance functions of the process. In this case, we do not have to know the sample paths of X.

A strong or weak solution X of the Itô stochastic differential equation (3.7) is called a *diffusion*. In particular, taking $a(t, x) = 0$ and $b(t, x) = 1$ in (3.7), we see that Brownian motion is a diffusion process.

In what follows, we only consider strong solutions of Itô stochastic differential equations.

First we give sufficient conditions for the existence and uniqueness of such solutions. A proof of the following result can be found, for example, in Kloeden and Platen (1992), Section 4.5, or in Øksendahl (1985), Theorem 5.5.

Assume the initial condition X_0 has a finite second moment: $EX_0^2 < \infty$, and is independent of $(B_t, t \geq 0)$.

Assume that, for all $t \in [0, T]$ and $x, y \in \mathbb{R}$, the coefficient functions $a(t, x)$ and $b(t, x)$ satisfy the following conditions:

- They are continuous.

- They satisfy a *Lipschitz condition* with respect to the second variable:

$$|a(t, x) - a(t, y)| + |b(t, x) - b(t, y)| \leq K |x - y|.$$

Then the Itô stochastic differential equation (3.7) has a unique strong solution X on $[0, T]$.

Example 3.2.2 (Linear stochastic differential equation)
Consider the Itô stochastic differential equation

$$X_t = X_0 + \int_0^t (c_1 X_s + c_2) \, ds + \int_0^t (\sigma_1 X_s + \sigma_2) \, dB_s, \quad t \in [0, T], \quad (3.8)$$

for constants c_i and σ_i, $i = 1, 2$.

The above conditions are satisfied (check them!) for

$$a(t, x) = c_1 x + c_2 \quad \text{and} \quad b(t, x) = \sigma_1 x + \sigma_2. \quad (3.9)$$

An Itô stochastic differential equation (3.8) with linear (in x) coefficient functions $a(t, x)$ and $b(t, x)$ is called a *linear Itô stochastic differential equation*. In Section 3.3 we will give the solution of the general linear stochastic differential equation. By virtue of the above theory, linear stochastic differential equations have a unique strong solution on every interval $[0, T]$, whatever the choice of the constants c_i and σ_i. $\qquad\qquad\qquad\qquad\qquad\qquad\qquad\qquad\qquad\qquad\Box$

3.2.2 Solving Itô Stochastic Differential Equations by the Itô Lemma

In this section we solve some elementary Itô stochastic differential equations by using the Itô lemma. Clearly, the first candidates are linear stochastic differential equations. They are "simple" and have unique strong solutions on every finite interval $[0, T]$; see Example 3.2.2.

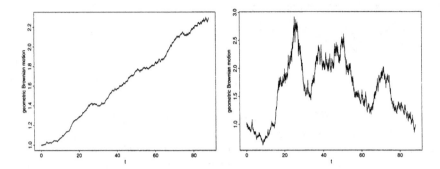

Figure 3.2.3 *Sample paths of geometric Brownian motion* $e^{(c-0.5\sigma^2)t+\sigma B_t}$ *with* $c = 0.01$. *Left:* $\sigma = 0.01$. *Right:* $\sigma = 0.1$.

Example 3.2.4 (Geometric Brownian motion as the solution of a linear Itô stochastic differential equation with multiplicative noise)
Consider the linear Itô stochastic differential equation

$$X_t = X_0 + c \int_0^t X_s \, ds + \sigma \int_0^t X_s \, dB_s \,, \quad t \in [0,T] \,, \tag{3.10}$$

for given constants c and $\sigma > 0$.

In the second, Itô, integral, Brownian motion and the process X are linked in a multiplicative way. Therefore the Itô stochastic differential equation (3.10) is also referred to as a linear Itô stochastic differential equation with *multiplicative noise*.

From (2.28) we know:

The particular geometric Brownian motion

$$X_t = X_0 \, e^{(c-0.5\sigma^2)t + \sigma B_t} \,, \quad t \in [0,T] \,, \tag{3.11}$$

solves (3.10), and from Example 3.2.2 we conclude that X is the unique solution of (3.10).

Figure 3.2.3 shows two paths of this process.

Recall how we verified that X satisfies (3.10): we applied the Itô lemma. Now we are faced with the inverse problem:

use the Itô lemma to find the solution (3.11) from (3.10).

Suppose $X_t = f(t, B_t)$ for some smooth function $f(t, x)$ and recall the Itô lemma from (2.26):

$$X_t = X_0 + \int_0^t \left[f_1(s, B_s) + \frac{1}{2} f_{22}(s, B_s) \right] ds + \int_0^t f_2(s, B_s) \, dB_s. \quad (3.12)$$

The process X is an Itô process, and therefore we may identify the integrands in the Riemann and Itô integrals, respectively, of (3.10) and (3.12) (see p. 119 for the necessary argument). This together with the continuity of the sample paths of Brownian motion yields the following system of two partial differential equations for f (because the partial derivatives of $f(t, x)$ are involved):

$$c f(t, x) = f_1(t, x) + \frac{1}{2} f_{22}(t, x), \quad (3.13)$$

$$\sigma f(t, x) = f_2(t, x). \quad (3.14)$$

From (3.14) we obtain

$$\sigma^2 f(t, x) = f_{22}(t, x).$$

Thus the two differential equations (3.13) and (3.14) can be simplified:

$$(c - 0.5 \sigma^2) f(t, x) = f_1(t, x), \quad \sigma f(t, x) = f_2(t, x). \quad (3.15)$$

Try to write $f(t, x)$ as a product of two functions:

$$f(t, x) = g(t) h(x).$$

Then (3.15) becomes

$$(c - 0.5 \sigma^2) g(t) = g'(t), \quad \sigma h(x) = h'(x).$$

Both of them can be solved by separation of variables (see Example 3.1.2):

$$g(t) = g(0) e^{(c - 0.5\sigma^2)t}, \quad h(x) = h(0) e^{\sigma x}.$$

Thus we obtain

$$f(t, x) = g(0) h(0) e^{(c - 0.5\sigma^2)t + \sigma x}.$$

Now recall that

$$X_0 = f(0, B_0) = f(0, 0) = g(0) h(0).$$

This finally gives

$$X_t = f(t, B_t) = X_0 e^{(c - 0.5 \sigma^2) t + \sigma B_t}, \quad t \in [0, T],$$

which agrees with (3.11).

Thus the solution to an Itô stochastic differential equation can sometimes be derived as the solution of a (deterministic) partial differential equation. □

Example 3.2.5 (The Ornstein–Uhlenbeck process)
We consider another linear stochastic differential equation:

$$
X_t = X_0 + c \int_0^t X_s \, ds + \sigma \int_0^t dB_s \,, \quad t \in [0,T] \,. \tag{3.16}
$$

Equation (3.16) is usually referred to as *Langevin equation*.

Langevin (1908) studied this kind of stochastic differential equation to model the velocity of a Brownian particle. This was long before a general theory for stochastic differential equations existed.

In contrast to (3.10), Brownian motion and the process X are not directly linked in the Itô integral part of (3.16). In the physics literature, the random forcing in (3.16) is called *additive noise* which is an adequate description of this phenomenon.

Model (3.16) is related to the world of time series analysis. Write (3.16) in the intuitive form

$$
dX_t = c \, X_t \, dt + \sigma \, dB_t \,,
$$

and *formally* set $dt = 1$. Then we have

$$
X_{t+1} - X_t = c \, X_t + \sigma \, (B_{t+1} - B_t)
$$

or

$$
X_{t+1} = \phi X_t + Z_t \,,
$$

where $\phi = c + 1$ is a constant and the random variables $Z_t = \sigma \, (B_{t+1} - B_t)$ constitute an iid sequence of $N(0, \sigma^2)$ random variables. This is an autoregressive process of order 1; cf. Example 1.2.3. This time series model can be considered as a discrete analogue of the solution to the Langevin equation (3.16).

To solve (3.16) the following transformation of X is convenient:

$$
Y_t = e^{-ct} X_t \,.
$$

Note that both processes X and Y satisfy the same initial condition:

$$
X_0 = Y_0 \,.
$$

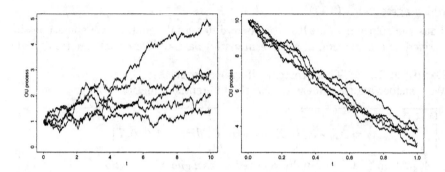

Figure 3.2.6 *Five sample paths of the Ornstein–Uhlenbeck process* (3.17). Left: $X_0 = 1$, $c = 0.1$, $\sigma = 1$. Right: $X_0 = 10$, $c = -1$, $\sigma = 1$.

Applying the Itô lemma (2.30) with

$$f(t,x) = e^{-ct}\, x\,, \quad f_1(t,x) = -c\, f(t,x)\,, \quad f_2(t,x) = e^{-ct}\,, \quad f_{22}(t,x) = 0\,,$$

and

$$A^{(1)} = c\, X \quad \text{and} \quad A^{(2)} = \sigma\,,$$

we obtain

$$
\begin{aligned}
Y_t - Y_0 &= \int_0^t \left[f_1(s, X_s) + c\, X_s\, f_2(s, X_s) + \frac{1}{2} \sigma^2\, f_{22}(s, X_s) \right] ds \\
&\quad + \int_0^t \left[\sigma\, f_2(s, X_s) \right] dB_s \\
&= \int_0^t \left[-c\, Y_s + c\, Y_s + 0 \right] ds\; +\; \int_0^t \left[\sigma\, e^{-cs} \right] dB_s \\
&= \int_0^t \left[\sigma\, e^{-cs} \right] dB_s\,.
\end{aligned}
$$

We conclude:

> The process
> $$X_t = e^{ct} X_0 \;+\; \sigma\, e^{ct} \int_0^t e^{-cs}\, dB_s\,. \qquad (3.17)$$
> solves the Langevin stochastic differential equation (3.16).

For a constant initial condition X_0, this process is called an *Ornstein–Uhlenbeck process*.

Figure 3.2.6 shows several paths of this process.

In order to verify that the process X, given by (3.17), is actually a solution to (3.16), apply the Itô lemma (2.30) to the process $X_t = u(t, Z_t)$, where

$$Z_t = \int_0^t e^{-cs} dB_s \quad \text{and} \quad u(t, z) = e^{ct} X_0 + \sigma e^{ct} z.$$

Moreover, since the Langevin equation is a linear Itô stochastic differential equation, we may conclude from Example 3.2.2 that X is the *unique* strong solution to (3.16).

We verify that the Ornstein–Uhlenbeck process is a Gaussian process. Assume for simplicity that $X_0 = 0$. Recall from the definition of the Itô stochastic integral that

$$\int_0^t e^{-cs} dB_s$$

is the mean square limit of approximating Riemann–Stieltjes sums

$$S_n = \sum_{i=1}^n e^{-c\, t_{i-1}} \left(B_{t_i} - B_{t_{i-1}} \right)$$

for partitions $\tau_n = (t_i)$ of $[0, t]$ with $\text{mesh}(\tau_n) \to 0$. The latter sum has a normal distribution with mean zero and variance

$$\sum_{i=1}^n e^{-2c\, t_{i-1}} \left(t_i - t_{i-1} \right). \tag{3.18}$$

(Check this!) Notice that (3.18) is a Riemann sum approximation to the integral

$$\int_0^t e^{-2c\, s} ds = \frac{1}{2c} \left(1 - e^{-2c\, t} \right).$$

Since mean square convergence implies convergence in distribution (see Appendix A1) we may conclude that the mean square limit X_t of the normally distributed Riemann–Stieltjes sums S_n is normally distributed. This follows from the argument given in Example A1.1 on p. 186. We have for $X_0 = 0$:

$$EX_t = 0 \quad \text{and} \quad \text{var}(X_t) = \frac{\sigma^2}{2c} \left(e^{2ct} - 1 \right).$$

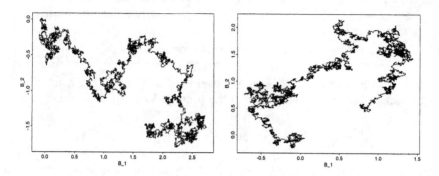

Figure 3.2.7 *Two paths of the two-dimensional process* $(B_t^{(1)}, B_t^{(2)})$, $t \in [0, T]$, *where* $B^{(1)}$ *and* $B^{(2)}$ *are two independent Brownian motions. See Example 3.2.8.*

Using the same Riemann–Stieltjes sum approach, you can also calculate the covariance function of an Ornstein–Uhlenbeck process with $X_0 = 0$:

$$\operatorname{cov}(X_s, X_t) = \frac{\sigma^2}{2c} \left(e^{c\,(t+s)} - e^{c\,(t-s)} \right), \quad s < t. \tag{3.19}$$

Since X is a mean-zero Gaussian process, this covariance function is characteristic for the Ornstein–Uhlenbeck process. $\qquad\square$

Example 3.2.8 (A stochastic differential equation with two independent driving Brownian motions)

Let $B^{(i)} = (B_t^{(i)}, t \geq 0)$ be two independent Brownian motions and σ_i, $i = 1, 2$, real numbers. See Figure 3.2.7 for an illustration of the two-dimensional process $(B_t^{(1)}, B_t^{(2)})$.

Define the process

$$\widetilde{B}_t = (\sigma_1^2 + \sigma_2^2)^{-1/2} \left(\sigma_1 B_t^{(1)} + \sigma_2 B_t^{(2)} \right).$$

Using the independence of $B^{(1)}$ and $B^{(2)}$, it is not difficult to see that

$$E\widetilde{B}_t = 0 \quad \text{and} \quad \operatorname{cov}(\widetilde{B}_t, \widetilde{B}_s) = \min(s, t),$$

i.e. $\widetilde{B}$ has exactly the same expectation and covariance functions as standard Brownian motion; see p. 35. Hence $\widetilde{B}$ is a Brownian motion.

Now consider the integral equation

$$X_t = X_0 + c \int_0^t X_s \, ds \; + \; \sigma_1 \int_0^t X_s \, dB_s^{(1)} \; + \; \sigma_2 \int_0^t X_s \, dB_s^{(2)} \,,$$

for constants c and σ_i.

We interpret the latter equation as

$$X_t - X_0 \;=\; c \int_0^t X_s \, ds \; + \; \int_0^t X_s \, d\left[\sigma_1 B_s^{(1)} \; + \; \sigma_2 B_s^{(2)}\right]$$

$$=\; c \int_0^t X_s \, ds \; + \; (\sigma_1^2 + \sigma_2^2)^{1/2} \int_0^t X_s \, d\widetilde{B}_s \,, \qquad (3.20)$$

which is an Itô stochastic differential equation with driving Brownian motion $\widetilde{B}$. From Example 3.2.4 we can read off the solution:

$$X_t \;=\; X_0 \, e^{[c-0.5\,(\sigma_1^2+\sigma_2^2)]\,t+(\sigma_1^2+\sigma_2^2)^{1/2}\widetilde{B}_t}$$

$$=\; X_0 \, e^{[c-0.5\,(\sigma_1^2+\sigma_2^2)]\,t+[\sigma_1 B_t^{(1)}+\sigma_2 B_t^{(2)}]} \,.$$

In the above definition (3.20) of the stochastic integral we were quite lucky because X appears as a multiplier in both integrals. Following similar patterns as for the definition of the Itô stochastic integral, it is also possible to introduce the stochastic integral

$$\int_0^t A_s^{(1)} \, dB_s^{(1)} \; + \; \int_0^t A_s^{(2)} \, dB_s^{(2)}$$

for more general processes $A^{(i)}$. Moreover, the Brownian motions $B^{(i)}$ can be dependent, and it is also possible to consider more than two driving Brownian motions.   □

3.2.3   Solving Itô Differential Equations via Stratonovich Calculus

In Section 2.4 we introduced the notion of Stratonovich stochastic integral. We also explained that some of the basic rules of classical calculus *formally* remain valid, in particular one can use the classical chain rule. We will exploit this property to solve some Itô stochastic differential equations.

Analogously to an Itô stochastic differential equation, a *Stratonovich stochastic differential equation* is a stochastic integral equation

$$X_t = X_0 \; + \; \int_0^t \widetilde{a}(s, X_s) \, ds \; + \; \int_0^t \widetilde{b}(s, X_s) \circ dB_s \,, \quad t \in [0,T], \qquad (3.21)$$

for given coefficient functions $\tilde{a}(t,x)$ and $\tilde{b}(t,x)$. The first integral is a Riemann integral, and the second one is a Stratonovich stochastic integral. As usual, $B = (B_t, t \geq 0)$ denotes Brownian motion. A stochastic process X is called a *solution* if it obeys (3.21). It is common to write (3.21) in the language of differentials:

$$dX_t = \tilde{a}(t, X_t)\, dt + \tilde{b}(t, X_t) \circ dB_t\,.$$

Now assume that X is the solution of the Itô stochastic differential equation

$$X_t = X_0 + \int_0^t a(s, X_s)\, ds + \int_0^t b(s, X_s)\, dB_s\,, \quad t \in [0, T]\,,$$

(3.22)

where the coefficient functions $a(t, x)$ and $b(t, x)$ satisfy the existence and uniqueness conditions on p. 138.

Recall the transformation formula (2.44) for Stratonovich integrals in terms of Itô and Riemann integrals:

$$\int_0^t f(s, X_s) \circ dB_s = \int_0^t f(s, X_s)\, dB_s + \frac{1}{2} \int_0^t b(s, X_s)\, f_2(s, X_s)\, ds\,.$$

(3.23)

For $f = b$ we obtain

$$\int_0^t b(s, X_s)\, dB_s = -\frac{1}{2} \int_0^t b(s, X_s)\, b_2(s, X_s)\, ds + \int_0^t b(s, X_s) \circ dB_s\,.$$

(3.24)

Plugging this relation into (3.22), we arrive at the Stratonovich stochastic differential equation:

$$X_t = X_0 + \int_0^t \tilde{a}(s, X_s)\, ds + \int_0^t b(s, X_s) \circ dB_s\,, \qquad (3.25)$$

where

$$\tilde{a}(t, x) = a(t, x) - \frac{1}{2} b(t, x)\, b_2(t, x)\,.$$

The Itô stochastic differential equation (3.22) is equivalent to the Stratonovich stochastic differential equation (3.25) in the sense that both have the same strong solution provided it exists for one of them.

Consider the stochastic process $Y_t = u(t, X_t)$, where X is the solution to (3.22) (equivalently, to (3.25)) and $u(t, x)$ is a smooth function. An application of the Itô lemma (2.30) with $A_t^{(1)} = a(t, X_t)$ and $A_t^{(2)} = b(t, X_t)$ yields

$$Y_t = Y_0 + \int_0^t \left[u_1 + a\,u_2 + \frac{1}{2}\,b^2\,u_{22} \right] ds + \int_0^t b u_2\,dB_s. \tag{3.26}$$

The functions a, b, u and their derivatives are clearly functions of s and X_s, but we suppress this dependence, in the notation, in order to make the above formula more compact. An application of the transformation formula (3.23) for $f = b u_2$ and $f_2 = b_2 u_2 + b u_{22}$ yields

$$\int_0^t b u_2\,dB_s = \int_0^t b u_2 \circ dB_s - \frac{1}{2}\int_0^t [b_2 u_2 + b u_{22}]\,b\,ds.$$

Combining this relation with (3.26), we finally obtain:

$$Y_t = Y_0 + \int_0^t [u_1 + (a - 0.5 b b_2)\,u_2]\,ds + \int_0^t b u_2 \circ dB_s$$

$$= Y_0 + \int_0^t [u_1 + \tilde{a} u_2]\,ds + \int_0^t b u_2 \circ dB_s. \tag{3.27}$$

This formula is the exact analogue of the classical chain rule for a twice differentiable function $u(t, x)$.

Indeed, assume for the moment that $x(t)$ satisfies the deterministic differential equation

$$dx(t) = \tilde{a}(t, x(t))\,dt + b(t, x(t))\,dc(t),$$

where $c(t)$ is a differentiable function. Then the classical rules of differentiation give the following formula:

$$u(t + dt, x + dx) - u(t, x)$$

$$= u_1(t, x)\,dt + u_2(t, x)\,dx$$

$$= [u_1(t, x) + \tilde{a}(t, x)\,u_2(t, x)]\,dt + b(t, x)\,u_2(t, x)\,dc.$$

This formal analogy with (3.27) allows one to solve Stratonovich stochastic differential equations by using the rules of classical calculus.

In what follows, we give some examples.

Example 3.2.9 (Solving an Itô stochastic differential equation by solving an equivalent Stratonovich stochastic differential equation)
Let f be a differentiable function.
A) Consider the Itô stochastic differential equation

$$X_t = X_0 + \frac{1}{2} \int_0^t f(X_s) f'(X_s)\, ds + \int_0^t f(X_s)\, dB_s. \qquad (3.28)$$

Thus

$$a(t,x) = \frac{1}{2} f(x) f'(x) \quad \text{and} \quad b(t,x) = f(x).$$

The corresponding Stratonovich stochastic differential equation (see (3.25)) with coefficient functions $\tilde{a}(t,x) = 0$ and $b(t,x)$ is then given by

$$X_t = X_0 + \int_0^t f(X_s) \circ dB_s.$$

It corresponds to the deterministic differential equation

$$dx(t) = f(x(t))\, dc(t),$$

where $c(t)$ is a differentiable function. Using classical separation of variables (see Example 3.1.2), we obtain

$$\int_{x(0)}^{x(t)} \frac{dx}{f(x)} = g(x(t)) - g(x(0)) = \int_0^t dc(s) = c(t) - c(0)$$

for some function $g(x)$. Then replace $x(t)$ with X_t and $c(t)$ with B_t:

$$g(X_t) - g(X_0) = B_t.$$

Hopefully, you can solve this equation for X to get an explicit solution of (3.28).

Now solve the Itô stochastic differential equation

$$X_t = X_0 + \frac{1}{2} n \int_0^t X_s^{2n-1}\, ds + \int_0^t X_s^n\, dB_s$$

for a positive integer $n > 1$.

B) A slightly more complicated Itô stochastic differential equation is given by

$$X_t - X_0 \tag{3.29}$$
$$= \int_0^t \left[q\, f(X_s) + \frac{1}{2} f(X_s)\, f'(X_s) \right] ds + \int_0^t f(X_s)\, dB_s$$

for constant q.

Notice that

$$a(t, x) = q\, f(x) + \frac{1}{2}\, f(x)\, f'(x) \quad \text{and} \quad b(t, x) = f(x).$$

The equivalent Stratonovich stochastic differential equation is

$$X_t = X_0 + \int_0^t [q\, f(X_s)]\, ds + \int_0^t f(X_s) \circ dB_s. \tag{3.30}$$

This corresponds to the deterministic differential equation

$$dx(t) = q\, f(x(t))\, dt + f(x(t))\, dc(t), \tag{3.31}$$

where $c(t)$ is a differentiable function. You can use the principle of separation of variables to solve this equation:

$$\int_{x(0)}^{x(t)} \frac{dx}{f(x)} = g(x(t)) - g(x(0)) = \int_0^t q\, ds + \int_0^t dc(s) = qt + c(t) - c(0)$$

for some function $g(x)$. You can check the validity of this approach by differentiating the integrals on the right- and left-hand sides: you obtain (3.31). An appeal to the Stratonovich chain rule (3.27) gives us the solution to (3.30), hence to (3.29), by replacing $x(t)$ with X_t and $c(t)$ with B_t:

$$g(X_t) - g(X_0) = qt + B_t.$$

Now consider the following example $f(x) = x + 1$ for $x \geq 0$ and $X_0 = 0$. Then $g(x) = \ln(1 + x)$ and $\ln(1 + X_t) = qt + B_t$, i.e.

$$X_t = -1 + e^{qt + B_t}. \qquad \square$$

Notes and Comments

The solution of stochastic differential equations is treated in all textbooks which are related to Itô calculus; see for example the references on p. 112.

In applications of stochastic differential equations one has to make a decision as to which kind of integral, Itô or Stratonovich, is more appropriate. There is no clear answer as to which type of differential equation should be used. Both kinds of equations are mathematically meaningful. The answer depends on how precisely we intend the noise process in the differential equation to model the real noise. Classical calculus is based on classes of smooth functions, in particular on differentiable functions. Real processes are often smooth with at least a small degree of correlation. This means that the noise is "sufficiently regular". In this case it seems reasonable to model the process by a Stratonovich stochastic differential equation because the rules of classical calculus remain valid.

3.3   The General Linear Differential Equation

Consider

The General Linear Stochastic Differential Equation:

$$X_t = X_0 + \int_0^t [c_1(s)X_s + c_2(s)]\, ds + \int_0^t [\sigma_1(s)X_s + \sigma_2(s)]\, dB_s\,, \quad t \in [0, T]\,.$$
$$(3.32)$$

The (deterministic) coefficient functions c_i and σ_i are continuous, hence bounded on $[0, T]$, and so it is not difficult to see that the existence and uniqueness conditions on p. 138 guarantee that (3.32) has a *unique strong solution.*

This equation is important for many applications. It is particularly attractive because it has an explicit solution in terms of the coefficient functions and of the underlying Brownian sample path. In what follows, we derive this solution by multiple use of different variants of the Itô lemma.

3.3.1   Linear Equations with Additive Noise

Setting $\sigma_1(t) = 0$ in (3.32), we obtain

The Linear Equation with Additive Noise:

$$X_t = X_0 + \int_0^t [c_1(s)X_s + c_2(s)]\, ds + \int_0^t \sigma_2(s)\, dB_s\,, \quad t \in [0, T]. \quad (3.33)$$

In terms of differentials it can be written as

$$dX_t = [c_1(t)X_t + c_2(t)]\, dt + \sigma_2(t)\, dB_t\,, \quad t \in [0, T].$$

The process X is not directly involved in the stochastic integral, and therefore (3.33) gained its name. The solution of (3.33) is particularly simple.

One way to solve a differential equation is to guess its form from some particular examples. This is perhaps not the most satisfactory approach since one certainly needs a lot of experience for such a guess. Anyway, let us assume that

$$X_t = [y(t)]^{-1} Y_t\,,$$

where

$$y(t) = \exp\left\{-\int_0^t c_1(s)\, ds\right\} \quad \text{and} \quad Y_t = f(t, X_t)\,,$$

for some smooth function $f(t, x)$. An application of the Itô lemma (2.30) on p. 120 yields

$$dY_t = d(y(t)X_t) = c_2(t)y(t)\, dt + \sigma_2(t)y(t)\, dB_t\,.$$

Integrating both sides and noticing that $y(0) = 1$, hence $X_0 = Y_0$, we obtain

---

The Solution of (3.33):

$$X_t = [y(t)]^{-1}\left(X_0 + \int_0^t c_2(s)y(s)\, ds + \int_0^t \sigma_2(s)y(s)\, dB_s\right). \quad (3.34)$$

If X_0 is a constant, this defines a Gaussian process. (You can check this by observing that the stochastic integral is the mean square limit, hence the limit in distribution, of Riemann–Stieltjes sums which are Gaussian random variables since the integrand $\sigma_2(s)y(s)$ is deterministic; cf. Example A1.1 on p. 186.)

Example 3.3.1 (The Langevin equation)

In Example 3.2.5 we already considered the Langevin equation. This is equation (3.33) with $c_1(t) = c$, $c_2(t) = 0$ and $\sigma_2(t) = \sigma$ for constants c and σ. The solution is given by

$$X_t = e^{ct}X_0 + \sigma e^{ct}\int_0^t e^{-cs}\,dB_s\,,\quad t \in [0,T]\,.$$

We learnt in Example 3.2.5 that X is called an Ornstein–Uhlenbeck process provided X_0 is constant. The covariance function of this Gaussian process is given in (3.19). $\square$

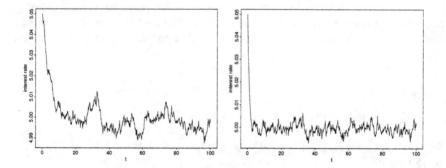

Figure 3.3.2 *Two sample paths of the Vasicek interest rate model* $dr_t = c\,[\mu - r_t]\,dt + \sigma\,dB_t$ *with* $\sigma = 0.5$, $r_0 = 5.05$, $\mu = 5$. *Left:* $c = 0.2$. *Right:* $c = 1$. *Both graphs show that* r_t *reverts from the initial value* $r_0 = 5.05$ *to the mean value* $\mu = 5$. *The speed at which this happens is determined by* c.

Example 3.3.3 (The Vasicek interest rate model)

This is one of the standard models for describing the interest rate for borrowing and lending money when this rate is not assumed to be a constant, but a random function r_t of time t. Since r may change at every instant of time, r_t is also called the *instantaneous interest rate*. In the *Vasicek model*, it is given by the linear stochastic differential equation

$$r_t = r_0 + c\int_0^t [\mu - r_s]\,ds + \sigma\int_0^t dB_s\,,\quad t \in [0,T]\,,\qquad (3.35)$$

or in the language of differentials:

$$dr_t = c\,[\mu - r_t]\,dt + \sigma\,dB_t\,,\quad t \in [0,T]\,,$$

where c, μ and σ are positive constants. The rationale of this model is that r_t fluctuates around the value μ. When r_t deviates from μ it will immediately be drawn back to μ; one says that r_t *reverts to the mean* μ. The speed at which this happens is adjusted by the parameter c. This point is illustrated well in Figure 3.3.2. This can also be seen from the form of the expectation and variance functions; see (3.36) and (3.37). The third parameter σ is a measure for the order of the magnitude of the fluctuations of r_t around μ; see (3.37). It is called *volatility*.

From the general solution (3.34) we can read off the solution to (3.35):

$$r_t = r_0 e^{-ct} + \mu \left(1 - e^{-ct}\right) + \sigma\, e^{-ct} \int_0^t e^{cs} dB_s \,.$$

We assume that r_0 is a constant. Then r is a Gaussian process. For $\mu = 0$ we obtain an Ornstein–Uhlenbeck process. It is not difficult to calculate

$$Er_t = r_0 e^{-ct} + \mu \left(1 - e^{-ct}\right) \tag{3.36}$$

and

$$\text{var}(r_t) = \frac{\sigma^2}{2c} \left(1 - e^{-2ct}\right). \tag{3.37}$$

You can check these formulae by using the results for the Ornstein–Uhlenbeck process; see Example 3.2.5. Alternatively, you can use the results in Section 3.3.4.

As $t \to \infty$, the random variable r_t converges in distribution to an $N(\mu, \sigma^2/(2c))$ variable. Since r_t is Gaussian, it assumes negative values with positive probability. This property is not very desirable for an interest rate. Nevertheless, if t is large and $\sigma^2/(2c)$ is small compared with μ, r_t is unlikely to be negative; see Figure 3.3.2 for an illustration of sample paths of the Vasicek process. There are various other models for interest rates which overcome this pathological property of the Vasicek process; see for example Lamberton and Lapeyre (1996). □

3.3.2   Homogeneous Equations with Multiplicative Noise

Another special case of the general linear stochastic differential equation (3.32) is given by

---

The Homogeneous Linear Equation:

$$X_t = X_0 + \int_0^t c_1(s)\, X_s\, ds + \int_0^t \sigma_1(s) X_s\, dB_s\,, \quad t \in [0,T], \quad (3.38)$$

or in the language of differentials,

$$dX_t = c_1(t)\, X_t\, dt + \sigma_1(t) X_t\, dB_t\,, \quad t \in [0,T]\,.$$

Since X_t appears as a factor of the increments of Brownian motion, (3.38) is called a stochastic differential equation with *multiplicative noise*. It is called *homogeneous* because $c_2(t) = \sigma_2(t) = 0$ for all t.

Since we may divide both sides of (3.38) by X_0 then we may assume that $X_0 = 1$. (The case $X_0 = 0$ corresponds to the trivial case $X_t = 0$ for all t which is not of interest.) Since we expect an exponential form of the solution, we assume $X_t > 0$ for all t. This allows one to consider $Y_t = \ln X_t = f(X_t)$ and to apply the Itô lemma (2.30) on p. 120 with

$$f(t,x) = \ln x\,, \quad f_1(t,x) = 0\,, \quad f_2(t,x) = x^{-1}\,, \quad f_{22}(t,x) = -x^{-2}\,.$$

We obtain

$$dY_t = [c_1(t) - 0.5\,\sigma_1^2(t)]\, dt + \sigma_1(t)\, dB_t\,. \quad (3.39)$$

First integrate both sides of this equation, then take exponentials and finally correct for the initial value if $X_0 \neq 1$:

The Solution of the Homogeneous Equation (3.38):

$$X_t = X_0 \exp\left\{ \int_0^t [c_1(s) - 0.5\,\sigma_1^2(s)]\, ds + \int_0^t \sigma_1(s)\, dB_s \right\},$$

$$t \in [0,T]\,. \quad (3.40)$$

You can check that X is a solution (do it!) by applying the Itô lemma to $f(Y_t) = X_0 \exp\{Y_t\}$, where dY_t is given by (3.39).

Example 3.3.4 (Geometric Brownian motion)
The most prominent example of a homogeneous stochastic differential equation

with multiplicative noise was treated in Example 3.2.4: $c_1(t) = c$, $\sigma_1(t) = \sigma$ for constants c and σ. We discovered that the geometric Brownian motion

$$X_t = X_0\, e^{(c-0.5\sigma^2)\,t+\sigma B_t}\,, \quad t \in [0,T]\,,$$

is the unique strong solution in this case. This agrees with the solution (3.40) of the general homogeneous stochastic differential equation (3.38). Geometric Brownian motion is of major importance for applications in finance; see Chapter 4.   □

3.3.3   The General Case

Now we are well prepared to solve the general linear stochastic differential equation (3.32). We achieve this by embedding (3.32) in a system of two stochastic differential equations.

Recall the solution Y of the homogeneous stochastic differential equation (3.38) with $Y_0 = 1$. It is given by formula (3.40) (with X replaced by Y). Consider the two processes

$$X_t^{(1)} = Y_t^{-1} \quad \text{and} \quad X_t^{(2)} = X_t\,.$$

Apply the Itô lemma (2.30) on p. 120 to Y_t^{-1} (with $f(t,x) = x^{-1}$). After some calculations we obtain

$$dX_t^{(1)} = [-c_1(t) + \sigma_1^2(t)]\,X_t^{(1)}\,dt - \sigma_1(t)\,X_t^{(1)}\,dB_t\,.$$

An appeal to the integration by parts formula (2.35) yields

$$d(X_t^{(1)}X_t^{(2)}) = [c_2(t) - \sigma_1(t)\sigma_2(t)]\,X_t^{(1)}\,dt + \sigma_2(t)\,X_t^{(1)}\,dB_t\,.$$

Now integrate both sides and recall that $Y_0 = 1$. Having in mind the particular forms of $X_t^{(1)}$ and $X_t^{(2)}$, we finally arrive at

---

The Solution of the General Linear Equation (3.32):

$$X_t \;=\; Y_t\left(X_0 + \int_0^t [c_2(t) - \sigma_1(t)\sigma_2(t)]Y_s^{-1}\,ds + \int_0^t \sigma_2(s)Y_s^{-1}\,dB_s\right),$$

$$t \in [0,T]\,. \tag{3.41}$$

> Here the process Y is the solution (3.40) of the homogeneous equation (3.38) (where X has to be replaced with Y) with the initial condition $Y_0 = 1$.

As a matter of fact, we also derived the solution of the corresponding deterministic differential equation

$$dx(t) = [c_1(t)x(t) + c_2(t)] \, dt \,, \quad t \in [0, T] \,.$$

Simply set $\sigma_1(t) = \sigma_2(t) = 0$ for all t. We will use this fact for calculating the expectation and second moment functions of X in Section 3.3.4.

3.3.4 The Expectation and Variance Functions of the Solution

If we have the explicit solution of a stochastic differential equation we can *in principle* determine its expectation, covariance and variance functions. We were able to proceed in this way for the Ornstein–Uhlenbeck process (see (3.19) on p. 144) and for geometric Brownian motion (see (1.16) and (1.17) on p. 42).

In what follows we go another way to calculate

$$\mu_X(t) = EX_t \quad \text{and} \quad q_X(t) = EX^2(t) \,, \quad t \in [0, T] \,.$$

Clearly, then we can also calculate $\sigma_X^2(t) = \text{var}(X_t)$.

Recall the general linear stochastic differential equation:

$$X_t = X_0 + \int_0^t [c_1(s)X_s + c_2(s)] \, ds + \int_0^t [\sigma_1(s)X_s + \sigma_2(s)] \, dB_s \,, \quad t \in [0, T] \,.$$

Take expectations on both sides and notice that the stochastic integral has expectation zero; see p. 111. Hence

$$\mu_X(t) = \mu_X(0) + \int_0^t [c_1(s) \, \mu_X(s) + c_2(s)] \, ds \,.$$

This corresponds to the general linear differential equation for $\mu_X(t)$

$$\mu_X'(t) = c_1(t) \, \mu_X(t) + c_2(t) \,.$$

Bearing in mind the remark at the end of the previous section, we can write down the solution μ_X of this differential equation in terms of c_1 and c_2. Derive

μ_X for the Ornstein–Uhlenbeck process and for geometric Brownian motion and compare with the known formulae.

Similarly one can proceed for $q_X(t)$. To obtain a stochastic differential equation for X_t^2, first apply the Itô lemma (2.30) on p. 120:

$$X_t^2 = X_0^2 + \int_0^t \left[2X_s(c_1(s)X_s + c_2(s)) + [\sigma_1(s)X_s + \sigma_2(s)]^2 \right] ds$$

$$+ \int_0^t \left[2X_s(\sigma_1(s)X_s + \sigma_2(s)) \right] dB_s \,,$$

then take expectations:

$$q_X(t) = q_X(0) + \int_0^t \left[[2c_1(s) + \sigma_1^2(s)] q_X(s) + 2[c_2(s) + \sigma_1(s)\sigma_2(s)] \mu_X(s) \right.$$

$$\left. + \sigma_2^2(s) \right] ds \,.$$

This is a general linear differential equation for $q_X(t)$ (notice that $\mu_X(t)$ is known):

$$q_X'(t) = [2c_1(t) + \sigma_1^2(t)]q_X(t) + 2[c_2(t) + \sigma_1(t)\sigma_2(t)] \mu_X(t) + \sigma_2^2(t) \,,$$

whose solution can be derived as a special case of (3.41); see the remark at the end of the previous section. No doubt: the solutions μ_X and q_X look in general quite messy, but in the case of constant coefficient functions c_i and σ_i you might try to obtain the expectation and variance functions of X.

3.4 Numerical Solution

Stochastic differential equations which admit an explicit solution are the exception from the rule. Therefore numerical techniques for the approximation of the solution to a stochastic differential equation are called for. In what follows, such an approximation is called a *numerical solution*.

Numerical solutions are needed for different aims. One purpose is to visualize a variety of sample paths of the solution. A collection of such paths is sometimes called a *scenario*. It gives an impression of the possible sample path behavior. In this sense, we can get some kind of "prediction" of the stochastic process at future instants of time. But a scenario has to be interpreted with care. In real life we never know the Brownian sample path driving the stochastic differential equation, and the simulation of a couple of such paths is not representative for the general picture.

A second objective (perhaps the most important one) is to achieve reasonable approximations to the distributional characteristics of the solution to a stochastic differential equation. They include expectations, variances, covariances and higher-order moments. This is indeed an important matter since only in a few cases one is able to give explicit formulae for these quantities, and even then they frequently involve special functions which have to be approximated numerically. Numerical solutions allow us to simulate as many sample paths as we want; they constitute the basis for Monte–Carlo techniques to obtain the distributional characteristics.

For the purpose of illustration we restrict ourselves to the numerical solution of the stochastic differential equation

$$dX_t = a(X_t)\,dt + b(X_t)\,dB_t\,, \quad t \in [0, T]\,, \tag{3.42}$$

where, as usual, B denotes Brownian motion, and the "differential equation" is actually a stochastic integral equation. We also assume that the coefficient functions $a(x)$ and $b(x)$ are Lipschitz continuous. If in addition $EX_0^2 < \infty$, these assumptions guarantee the existence and uniqueness of a strong solution; see p. 138.

3.4.1   The Euler Approximation

A *numerical solution* $X^{(n)} = (X_t^{(n)}, t \in [0, T])$ of the stochastic differential equation (3.42) is a stochastic process that approximates the solution $X = (X_t, t \in [0, T])$ of (3.42) in a sense to be made precise later. Such a solution is characterized by a partition τ_n of $[0, T]$:

$$\tau_n : \quad 0 = t_0 < t_1 < \cdots < t_{n-1} < t_n = T\,,$$

with mesh

$$\delta_n = \text{mesh}(\tau_n) = \max_{i=1,\ldots,n} (t_i - t_{i-1}) = \max_{i=1,\ldots,n} \Delta_i\,.$$

The process $X^{(n)}$ is calculated only at the points t_i of the partition τ_n, and so one has some freedom to choose $X^{(n)}$ on the intervals $[t_{i-1}, t_i]$. Since we are interested in solutions X with continuous sample paths we will assume that $X_t^{(n)}$ on (t_{i-1}, t_i) is obtained by simple linear interpolation of the points $(t_{i-1}, X_{t_{i-1}}^{(n)})$ and $(t_i, X_{t_i}^{(n)})$.

A numerical approximation scheme determines $X^{(n)}$ at the points t_i. A naive way to obtain a numerical solution is to replace the differentials in the stochastic differential equation (3.42) with differences. This leads to the following iterative scheme:

The Euler Approximation Scheme:

Denote, as usual,

$$\Delta_i = t_i - t_{i-1} \quad \text{and} \quad \Delta_i B = B_{t_i} - B_{t_{i-1}}, \quad i = 1, \ldots, n.$$

Then

$$
\begin{aligned}
X_0^{(n)} &= X_0, \\
X_{t_1}^{(n)} &= X_0^{(n)} + a(X_0^{(n)})\Delta_1 + b(X_0^{(n)})\Delta_1 B, \\
X_{t_2}^{(n)} &= X_{t_1}^{(n)} + a(X_{t_1}^{(n)})\Delta_2 + b(X_{t_1}^{(n)})\Delta_2 B, \\
&\ \ \vdots \\
X_{t_{n-1}}^{(n)} &= X_{t_{n-2}}^{(n)} + a(X_{t_{n-2}}^{(n)})\Delta_{n-1} + b(X_{t_{n-2}}^{(n)})\Delta_{n-1} B, \\
X_T^{(n)} &= X_{t_{n-1}}^{(n)} + a(X_{t_{n-1}}^{(n)})\Delta_n + b(X_{t_{n-1}}^{(n)})\Delta_n B.
\end{aligned}
$$

In practice one usually chooses equidistant points t_i such that

$$\delta_n = \mathrm{mesh}(\tau_n) = T/n,$$

and

$$X_{iT/n}^{(n)} = X_{(i-1)T/n}^{(n)} + a(X_{(i-1)T/n}^{(n)})\,\delta_n + b(X_{(i-1)T/n}^{(n)})\,\Delta_i B,$$

$$i = 1, \ldots, n. \tag{3.43}$$

Figures 3.4.1 and 3.4.2 show how the Euler approximation works in practice. Several questions naturally arise:

(a) *In which sense is the numerical solution $X^{(n)}$ close to the solution X?*

(b) *How can we measure the quality of the approximation of X by $X^{(n)}$?*

(c) *Is there an approximation to X which is better than the Euler scheme?*

The third question has a positive answer; see Section 3.4.2. The first and second questions have several answers. If we are interested in the pathwise approximation of X, we would like that $X^{(n)}(\omega)$ is "close" to the sample path $X(\omega)$ for a given Brownian path $B(\omega)$. In practice we never know the latter

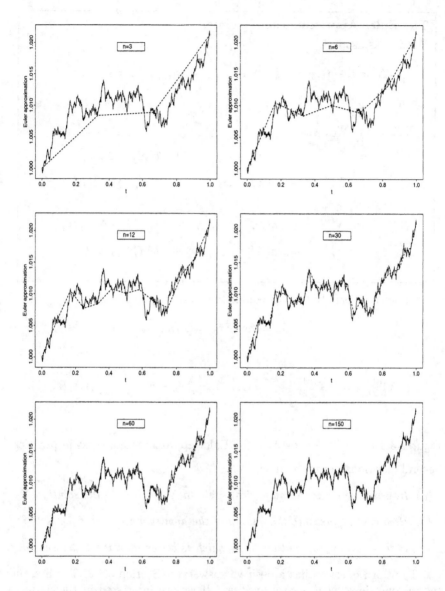

Figure 3.4.1 The equidistant Euler scheme (3.43) at work: *numerical solutions (dashed lines) and exact solution to the stochastic differential equation* $dX_t = 0.01X_t\,dt + 0.01X_t\,dB_t,\ t \in [0,1],\ with\ X_0 = 1.$

path, but we can easily simulate Brownian paths on the computer; see Section 1.3.3 for a brief introduction to this topic. As a measure of the quality of pathwise approximation one can choose the quantity

$$e_s(\delta_n) = E|X_T(\omega) - X_T^{(n)}(\omega)|\,.$$

Notice that e_s indeed describes the ω-wise (i.e. pathwise) comparison of X and $X^{(n)}$ at $t = T$. The index "s" in e_s stands for "strong"; a pathwise approximation of X is usually called a *strong numerical solution*; see below. There are certainly more appropriate criteria to describe the pathwise closeness of X and $X^{(n)}$; one of them could be $E \sup_{t \in [0,T]} |X_t(\omega) - X_t^{(n)}(\omega)|$. However, the latter quantity is more difficult to deal with theoretically. So let us believe that the paths $X^{(n)}(\omega)$ and $X(\omega)$ are close whenever they are close at the end of the interval $[0, T]$.

We say that $X^{(n)}$ is a *strong numerical solution* of the stochastic differential equation (3.42) if

$$e_s(\delta_n) \to 0 \quad \text{as} \quad \delta_n = \text{mesh}(\tau_n) \to 0\,.$$

In contrast to a strong numerical solution, a *weak numerical solution* aims at the approximation of the moments of the solution X. In this case it is not essential how close $X_t(\omega)$ and $X_t^{(n)}(\omega)$ really are. What matters is that the difference of moments

$$e_w(\delta_n) = \left| E f(X_T) - E f(X_T^{(n)}) \right|$$

is small. Here f is chosen from a class of smooth functions, for example certain polynomials or functions with a specific polynomial growth. We refrain from discussing these classes of functions because their definitions are technical, and they also differ in the literature. The "w" in e_w stands for "weak". As before, we may wonder why we compare the moments $E f(X_T)$ and $E f(X_T^{(n)})$ only for $t = T$, but again, this is for theoretical simplicity.

We say that $X^{(n)}$ is a *weak numerical solution* of the stochastic differential equation (3.42) if

$$e_w(\delta_n) \to 0 \quad \text{as} \quad \delta_n = \text{mesh}(\tau_n) \to 0\,.$$

In order to measure the quality of the approximation of X by $X^{(n)}$ one introduces the *order of convergence*:

The numerical solution $X^{(n)}$ *converges strongly to X with order* $\gamma > 0$ if there exists a constant $c > 0$ such that

$$e_s(\delta_n) \leq c \, \delta_n^\gamma \quad \text{for} \quad \delta_n \leq \delta_0 \,.$$

The numerical solution $X^{(n)}$ *converges weakly to X with order* $\gamma > 0$ if there exists a constant $c > 0$ such that

$$e_w(\delta_n) \leq c \, \delta_n^\gamma \quad \text{for} \quad \delta_n \leq \delta_0 \,.$$

The equidistant Euler approximation (see (3.43)) converges strongly with order 0.5 and weakly with order 1.0 (for a class of functions with appropriate polynomial growth).

3.4.2   The Milstein Approximation

The Euler approximation can be further improved. We illustrate this by introducing the *Milstein approximation*. As before, we consider the stochastic differential equation

$$X_t = X_0 + \int_0^t a(X_s)\,ds + \int_0^t b(X_s)\,dB_s \,, \quad t \in [0, T] \,.$$

For the points t_i of the partition τ_n we can consider the difference $X_{t_i} - X_{t_{i-1}}$ and obtain:

$$X_{t_i} = X_{t_{i-1}} + \int_{t_{i-1}}^{t_i} a(X_s)\,ds + \int_{t_{i-1}}^{t_i} b(X_s)\,dB_s \,, \quad i = 1, \ldots, n \,. \qquad (3.44)$$

The Euler approximation is based on a discretization of the integrals in (3.44). To see this, first consider the approximations

$$\int_{t_{i-1}}^{t_i} a(X_s)\,ds \approx a(X_{t_{i-1}})\,\Delta_i \,, \quad \int_{t_{i-1}}^{t_i} b(X_s)\,dB_s \approx b(X_{t_{i-1}})\,\Delta_i B \,,$$

and then replace X_{t_i} with $X_{t_i}^{(n)}$:

$$X_{t_i^{(n)}} = X_{t_{i-1}^{(n)}} + a(X_{t_{i-1}}^{(n)})\,\Delta_i + b(X_{t_{i-1}}^{(n)})\,\Delta_i B \,, \quad i = 1, \ldots, n \,.$$

The Milstein approximation exploits a so-called *Taylor–Itô expansion* of relation (3.44). The idea consists of applying the Itô lemma on p. 120 to the integrands $a(X_s)$ and $b(X_s)$ in (3.44). To make the formulae below more compact we write $a, b, a', \ldots$ instead of $a(X_y), b(X_y), a'(X_y), \ldots$.

$$X_{t_i} - X_{t_{i-1}}$$

$$= \int_{t_{i-1}}^{t_i} \left[a(X_{t_{i-1}}) + \int_{t_{i-1}}^s \left(aa' + \frac{1}{2} b^2 a'' \right) dy + \int_{t_{i-1}}^s ba' dB_y \right] ds$$

$$+ \int_{t_{i-1}}^{t_i} \left[b(X_{t_{i-1}}) + \int_{t_{i-1}}^s \left(ab' + \frac{1}{2} b^2 b'' \right) dy + \int_{t_{i-1}}^s bb' dB_y \right] dB_s$$

$$= a(X_{t_{i-1}}) \Delta_i + b(X_{t_{i-1}}) \Delta_i B + R_i \,, \tag{3.45}$$

where the remainder term is given by

$$R_i = R_i^{(1)} + R_i^{(2)} = \int_{t_{i-1}}^{t_i} \left[\int_{t_{i-1}}^s bb' dB_y \right] dB_s + R_i^{(2)} \,. \tag{3.46}$$

The double stochastic integral $R_i^{(1)}$ is then approximated by

$$R_i^{(1)} \approx b(X_{t_{i-1}}) b'(X_{t_{i-1}}) \int_{t_{i-1}}^{t_i} \left(\int_{t_{i-1}}^s dB_y \right) dB_s \,. \tag{3.47}$$

Denote the double integral on the right-hand side by I. The evaluation of I is quite delicate. Recall from p. 99 that $(dB_s)^2 = ds$, and therefore

$$\int_{t_{i-1}}^{t_i} (dB_s)^2 = \int_{t_{i-1}}^{t_i} ds = \Delta_i \,.$$

Consider the double integral

$$(\Delta_i B)^2 \tag{3.48}$$

$$= \left(\int_{t_{i-1}}^{t_i} dB_s \right) \left(\int_{t_{i-1}}^{t_i} dB_y \right) = \int_{t_{i-1}}^{t_i} \left(\int_{t_{i-1}}^{t_i} dB_y \right) dB_s \,.$$

Alternatively, we interpret the latter integral in a heuristic sense as

$$\int_{t_{i-1}}^{t_i} \left(\int_{t_{i-1}}^s dB_y \right) dB_s + \int_{t_{i-1}}^{t_i} \left(\int_s^{t_i} dB_y \right) dB_s + \int_{t_{i-1}}^{t_i} (dB_s)^2$$

$$= 2 \int_{t_{i-1}}^{t_i} \left(\int_{t_{i-1}}^{s} dB_y \right) dB_s + \Delta_i = 2I + \Delta_i \,. \tag{3.49}$$

Combining (3.47) with (3.48) and (3.49), we have

$$R_i^{(1)} \approx \frac{1}{2} b(X_{t_{i-1}}) b'(X_{t_{i-1}}) \left[(\Delta_i B)^2 - \Delta_i \right]. \tag{3.50}$$

Under mild assumptions on the coefficient functions $a(x)$ and $b(x)$ one can show that $R_i^{(2)}$ is small in comparison with $R_i^{(1)}$. Thus the latter term determines the order of magnitude of R_i. Relations (3.45) and (3.50) give the motivation for the definition of the *Milstein approximation scheme*. Notice that this scheme is the Euler approximation with an additional correction term containing the squared increments of Brownian motion.

The Milstein Approximation Scheme:
$X_0^{(n)} = X_0$ and for $i = 1, \ldots, n$,

$$X_{t_i}^{(n)} = X_{t_{i-1}}^{(n)} + a(X_{t_{i-1}}^{(n)}) \Delta_i + b(X_{t_{i-1}}^{(n)}) \Delta_i B$$

$$+ \frac{1}{2} b(X_{t_{i-1}}^{(n)}) \, b'(X_{t_{i-1}}^{(n)}) \left[(\Delta_i B)^2 - \Delta_i \right].$$

The Milstein approximation leads to a substantial improvement of the quality of the numerical solution:

The equidistant Milstein approximation converges strongly with order 1.0.

This improvement is illustrated well in Figure 3.4.2.

Notes and Comments

The Milstein approximation can be further improved by applying the Itô lemma to the integrands in the remainder term R_i; see (3.46). The form of the numerical solutions then becomes more and more complicated. However, in view of the power of modern computers, this is not an issue one has to worry about.

Taylor–Itô expansions such as (3.45) involve multiple stochastic integrals. Their rigorous treatment requires a more advanced level of the theory. This also concerns the derivation of (3.50); the latter relation was obtained by some heuristic arguments.

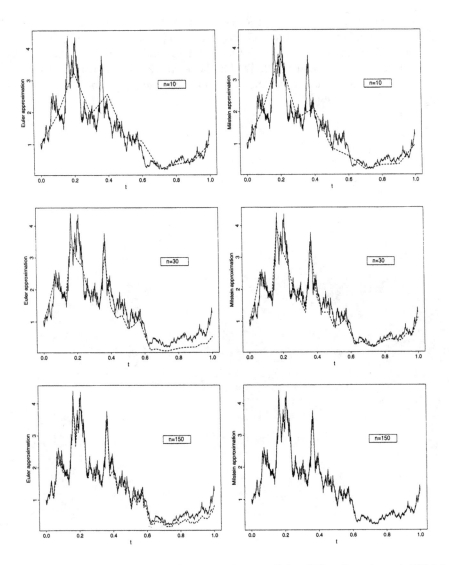

Figure 3.4.2 *A comparison of the equidistant Euler* (left column) *and Milstein* (right column) *schemes. In every figure a numerical solution (dashed lines) and the exact solution to the stochastic differential equation* $dX_t = 0.01X_t\,dt + 2X_t\,dB_t$, $t \in [0, 1]$, *with* $X_0 = 1$ *are given. Notice the significant improvement of the approximation by the Milstein scheme for large n, in particular at the end of the interval.*

The numerical solution of stochastic differential equations is a relatively new area of applied probability theory. An overview of the existing numerical techniques is given in Kloeden and Platen (1992) and the companion book by Kloeden, Platen and Schurz (1994). The latter book aims at the introduction to stochastic differential equations and their numerical solution via computer experiments.

4

Applications of Stochastic Calculus in Finance

Since the celebrated papers by Black and Scholes (1973) and Merton (1973) the idea of using stochastic calculus for modelling prices of risky assets (share prices of stock, stock indices such as the Dow Jones, Nikkei or DAX, foreign exchange rates, interest rates, etc.) has been generally accepted. This led to a new branch of applied probability theory, the field of mathematical finance. It is a symbiosis of stochastic modelling, economic reasoning and practical financial engineering.

In this chapter we consider the Black–Scholes model for pricing a European call option. Do not worry if you do not have much prior knowledge about economic and financial processes. As a minimum, you will need some words of economic language, and as a maximum, the Itô lemma. In Section 4.1 we will explain the basic terminology of finance: *bond, stock, option, portfolio, volatility, trading strategy, hedging, maturity of a contract, self-financing, arbitrage* will be the catchy words. The Black–Scholes formula for pricing a European call option will be obtained as the solution of a particular partial differential equation.

Since notions like *equivalent martingale measure* and *change of measure* have become predominant in the financial literature, we give in Section 4.2 a short introduction to this area. This requires some knowledge of measure-theoretic tools, but, as usual, we will keep the theory on a low level. We hope that the key ideas will become transparent, and we give a second derivation of the Black–Scholes formula by using the change of measure ideology.

4.1   The Black–Scholes Option Pricing Formula

4.1.1   A Short Excursion into Finance

We assume that the price X_t of a *risky asset* (called *stock*) at time t is given by geometric Brownian motion of the form

$$X_t = f(t, B_t) = X_0 \, e^{(c-0.5\,\sigma^2)\,t \, + \, \sigma \, B_t} \,, \tag{4.1}$$

where, as usual, $B = (B_t, t \geq 0)$ is Brownian motion, and X_0 is assumed to be independent of B. The motivation for this assumption on X comes from the fact that X is the unique strong solution of the linear stochastic differential equation

$$X_t = X_0 + c \int_0^t X_s \, ds \, + \, \sigma \int_0^t X_s \, dB_s \,, \tag{4.2}$$

which we can formally write as

$$dX_t = c\,X_t\,dt \, + \, \sigma\,X_t\,dB_t \,.$$

This was proved in Example 3.2.4. If we interpret this equation in a naive way, we have on $[t, t+dt]$:

$$X_{t+dt} - X_t = c\,X_t\,dt \, + \, \sigma\,X_t\,dB_t \,.$$

Equivalently,

$$\frac{X_{t+dt} - X_t}{X_t} = c\,dt \, + \, \sigma\,dB_t \,.$$

The quantity on the left-hand side is the *relative return* from the asset in the period of time $[t, t+dt]$. It tells us that there is a linear trend $c\,dt$ which is disturbed by a stochastic noise term $\sigma\,dB_t$. The constant $c > 0$ is the so-called *mean rate of return*, and $\sigma > 0$ is the *volatility*. A glance at formula (4.1) tells us that, the larger σ, the larger the fluctuations of X_t. You can also check this with the formula for the variance function of geometric Brownian motion, which is provided in (1.18). Thus σ is a measure of the riskiness of the asset.

It is believed that the model (4.2) is a reasonable, though crude, first approximation to a real price process. If you forget for the moment the term with σ, i.e. assume $\sigma = 0$, then (4.2) is a deterministic differential equation which has the well-known solution $X_t = X_0 \exp\{ct\}$. Thus, if $\sigma > 0$, we should expect to obtain a randomly perturbed exponential function, and this is the geometric Brownian motion (4.1). People in economics believe in exponential growth, and therefore they are quite satisfied with this model.

Now assume that you have a non-risky asset such as a bank account. In financial theory, it is called a *bond*. We assume that an investment of β_0 in bond yields an amount of

$$\beta_t = \beta_0 \, e^{rt}$$

at time t. Thus your initial capital β_0 has been continuously compounded with a constant interest rate $r > 0$. This is an idealization since the interest rate changes with time as well. Note that β satisfies the deterministic integral equation

$$\beta_t = \beta_0 \, + \, r \int_0^t \beta_s \, ds \,. \tag{4.3}$$

In general, you want to hold certain amounts of shares: a_t in stock and b_t in bond. They constitute your *portfolio*. We assume that a_t and b_t are stochastic processes adapted to Brownian motion and call the pair

$$(a_t, b_t) \,, \quad t \in [0, T] \,,$$

a *trading strategy*. Clearly, you want to choose a strategy, where you do not lose. How to choose (a_t, b_t) in a reasonable way, will be discussed below. Notice that your *wealth* V_t (or the *value of your portfolio*) at time t is now given by

$$V_t = a_t \, X_t + b_t \, \beta_t \,.$$

We allow both, a_t and b_t, to assume any positive or negative values. A negative value of a_t means *short sale* of stock, i.e. you sell the stock at time t. A negative value of b_t means that you *borrow* money at the bond's riskless interest rate r. In reality, you would have to pay *transaction costs* for operations on stock and sale, but we neglect them here for simplicity. Moreover, we do not assume that a_t and b_t are bounded. So, in principle, you should have a potentially infinite amount of capital, and you should allow for unbounded debts as well. Clearly, this is a simplification, which makes our mathematical problems easier. And finally, we assume that you spend no money on other purposes, i.e. you do not make your portfolio smaller by *consumption*.

We assume that your trading strategy (a_t, b_t) is *self-financing*. This means that the increments of your wealth V_t result only from changes of the prices X_t and β_t of your assets. We formulate the *self-financing condition* in terms of differentials:

$$dV_t = d(a_t X_t + b_t \beta_t) = a_t \, dX_t + b_t \, d\beta_t \,,$$

which we interpret in the Itô sense as the relation

$$V_t - V_0 = \int_0^t d(a_s \, X_s + b_s \, \beta_s) = \int_0^t a_s \, dX_s \, + \, \int_0^t b_s \, d\beta_s \,.$$

The integrals on the right-hand side clearly make sense if you replace dX_s with $cX_s\,ds + \sigma X_s\,dB_s$, see (4.2), and $d\beta_s$ with $r\beta_s\,ds$, see (4.3). Hence the value V_t of your portfolio at time t is precisely equal to the initial investment V_0 plus capital gains from stock and bond up to time t.

4.1.2   What is an Option?

Now suppose you purchase a ticket, called an *option*, at time $t = 0$ which entitles you to buy one share of stock until or at time T, the *time of maturity* or *time of expiration* of the option. If you can exercise this option at a fixed price K, called the *exercise price* or *strike price* of the option, *only* at time of maturity T, this is called a *European call option*. If you can exercise it until or at time T, it is called an *American call option*. Note that there are many more different kinds of options in the real world of finance but we will not be able to include them in this book.

The holder of a call option is *not* obliged to exercise it. Thus, if at time T the price X_T is less than K, the holder of the ticket would be silly to exercise it (you could buy one share for X_T\$ on the market!), and so the ticket expires as a worthless contract. If the price X_T exceeds K, it is worthwhile to exercise the call, i.e. one buys the share at the price K, then turns around and sells it at the price X_T for a net profit $X_T - K$.

In sum, the purchaser of a European call option is entitled to a payment of

$$(X_T - K)^+ = \max(0, X_T - K) = \begin{cases} X_T - K, & \text{if } X_T > K, \\ 0, & \text{if } X_T \le K. \end{cases}$$

See Figure 4.1.1 for an illustration.

A *put* is an option to sell stock at a given price K on or until a particular date of maturity T. A *European put option* is exercised *only* at time of maturity, an *American put* can be exercised until or at time T. The purchaser of a European put makes profit

$$(K - X_T)^+ = \begin{cases} K - X_T, & \text{if } X_T < K, \\ 0, & \text{if } X_T \ge K. \end{cases}$$

In our theoretical considerations we restrict ourselves to European calls. This has a simple reason: in this case we can derive explicit solutions and compact formulae for our pricing problems. Thus,

from now on, an option is a European call option.

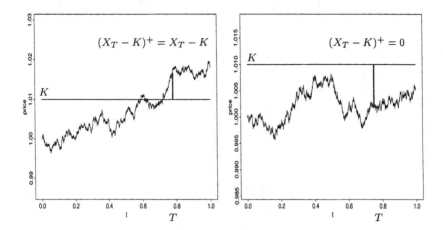

Figure 4.1.1 *The value of a European call option with exercise price K at time of maturity T.*

As an aside, it is interesting to note that the situation can be imagined colorfully as a game where the reward is the payoff of the option and the option holder pays a fee (the option price) for playing the game.

Since you do *not* know the price X_T at time $t = 0$, when you purchase the call, a natural question arises:

How much would you be willing to pay for such a ticket, i.e. what is a rational price for this option at time $t = 0$?

Black, Scholes and Merton defined such a value as follows:

- An individual, after investing this rational value of money in stock and bond at time $t = 0$, can manage his/her portfolio according to a self-financing strategy (see p. 169) so as to yield the same payoff $(X_T - K)^+$ as if the option had been purchased.

- If the option were offered at any price other than this rational value, there would be an opportunity of *arbitrage*, i.e. for unbounded profits without an accompanying risk of loss.

4.1.3 A Mathematical Formulation of the Option Pricing Problem

Now suppose we want to find a self-financing strategy (a_t, b_t) and an associated value process V_t such that

$$V_t = a_t X_t + b_t \beta_t = u(T - t, X_t), \quad t \in [0, T],$$

for some smooth deterministic function $u(t, x)$. Clearly, this is a restriction: you assume that the value V_t of your portfolio depends in a smooth way on t and X_t. It is our aim to find this function $u(t, x)$. Since the value V_T of the portfolio at time of maturity T shall be $(X_T - K)^+$, we get the *terminal condition*

$$V_T = u(0, X_T) = (X_T - K)^+. \tag{4.4}$$

In the financial literature, the process of building a self-financing strategy such that (4.4) holds is called *hedging against the contingent claim* $(X_T - K)^+$.

We intend to apply the Itô lemma to the value process $V_t = u(T - t, X_t)$. Write $f(t, x) = u(T - t, x)$ and notice that

$$f_1(t, x) = -u_1(T - t, x), \quad f_2(t, x) = u_2(T - t, x), \quad f_{22}(t, x) = u_{22}(T - t, x).$$

Also recall that X satisfies the Itô integral equation

$$X_t = X_0 + c \int_0^t X_s \, ds + \sigma \int_0^t X_s \, dB_s.$$

Now an application of the Itô lemma (2.30) with $A^{(1)} = c\,X$ and $A^{(2)} = \sigma\,X$ yields that

$$
\begin{aligned}
V_t - V_0 &= f(t, X_t) - f(0, X_0) \\[2mm]
&= \int_0^t \left[f_1(s, X_s) + c\,X_s\,f_2(s, X_s) + 0.5\,\sigma^2\,X_s^2\,f_{22}(s, X_s) \right] ds \\[2mm]
&\quad + \int_0^t \left[\sigma\,X_s\,f_2(s, X_s) \right] dB_s \\[2mm]
&= \int_0^t \left[-u_1(T - s, X_s) + c\,X_s\,u_2(T - s, X_s) \right. \\[2mm]
&\qquad\qquad \left. + 0.5\,\sigma^2\,X_s^2\,u_{22}(T - s, X_s) \right] ds \\[2mm]
&\quad + \int_0^t \left[\sigma\,X_s\,u_2(T - s, X_s) \right] dB_s. \tag{4.5}
\end{aligned}
$$

On the other hand, (a_t, b_t) is self-financing:

$$V_t - V_0 = \int_0^t a_s \, dX_s + \int_0^t b_s \, d\beta_s . \tag{4.6}$$

Since $\beta_t = \beta_0 e^{rt}$,

$$d\beta_t = r \, \beta_0 \, e^{rt} \, dt = r \, \beta_t \, dt . \tag{4.7}$$

Moreover, $V_t = a_t \, X_t + b_t \, \beta_t$, thus

$$b_t = \frac{V_t - a_t X_t}{\beta_t} . \tag{4.8}$$

Combining (4.6)–(4.8), we obtain another expression for

$$
\begin{aligned}
V_t - V_0 &= \int_0^t a_s \, dX_s + \int_0^t \frac{V_s - a_s X_s}{\beta_s} r \, \beta_s \, ds \\
&= \int_0^t a_s \, dX_s + \int_0^t r \, (V_s - a_s \, X_s) \, ds \\
&= \int_0^t c \, a_s \, X_s \, ds + \int_0^t \sigma \, a_s \, X_s \, dB_s + \int_0^t r \, (V_s - a_s \, X_s) \, ds \\
&= \int_0^t [(c - r) \, a_s \, X_s + r \, V_s] \, ds + \int_0^t [\sigma \, a_s \, X_s] \, dB_s .
\end{aligned}
\tag{4.9}
$$

Now compare formulae (4.5) and (4.9). We learnt on p. 119 that coefficient functions of Itô processes coincide. Thus we may formally identify the integrands of the Riemann and Itô integrals, respectively, in (4.5) and (4.9):

$$a_t = u_2(T - t, X_t), \tag{4.10}$$

$$
\begin{aligned}
(c - r) \, a_t \, X_t + r \, u(T - t, X_t) &= (c - r) \, u_2(T - t, X_t) \, X_t + r \, u(T - t, X_t) \\
&= -u_1(T - t, X_t) + c \, X_t \, u_2(T - t, X_t) \\
&\quad + 0.5 \, \sigma^2 \, X_t^2 \, u_{22}(T - t, X_t) .
\end{aligned}
$$

Since X_t may assume any positive value, we can write the last identity as a partial differential equation ("partial" refers to the use of the partial derivatives of u):

$$u_1(t, x) = 0.5 \, \sigma^2 \, x^2 \, u_{22}(t, x) + r \, x \, u_2(t, x) - r \, u(t, x), \tag{4.11}$$

$$x > 0, \, t \in [0, T].$$

Recalling the terminal condition (4.4), we also require that

$$V_T = u(0, X_T) = (X_T - K)^+ .$$

This yields the deterministic terminal condition

$$u(0, x) = (x - K)^+ , \quad x > 0 . \qquad (4.12)$$

4.1.4   The Black and Scholes Formula

In general, it is hard to solve a partial differential equation explicitly, and so one has to rely on numerical solutions. So it is somewhat surprising that the partial differential equation (4.11) has an explicit solution. (This is perhaps one of the reasons for the popularity of the Black–Scholes–Merton approach.) The partial differential equation (4.11) with terminal condition (4.12) has been well-studied; see for example Zauderer (1989). It has the explicit solution

$$u(t, x) = x\, \Phi(g(t, x)) - K\, e^{-rt}\, \Phi(h(t, x)) ,$$

where

$$g(t, x) \;\; = \;\; \frac{\ln(x/K) + (r + 0.5\, \sigma^2)\, t}{\sigma t^{1/2}} ,$$

$$h(t, x) \;\; = \;\; g(t, x) - \sigma\, t^{1/2} ,$$

and

$$\Phi(x) = \frac{1}{(2\pi)^{1/2}} \int_{-\infty}^{x} e^{-y^2/2}\, dy , \quad x \in \mathbb{R},$$

is the standard normal distribution function.

After all these calculations,

what did we actually gain?

Recalling our starting point on p. 171, we see that

$$V_0 = u(T, X_0) = X_0\, \Phi(g(T, X_0)) - K\, e^{-rT}\, \Phi(h(T, X_0)) \qquad (4.13)$$

is a rational price at time $t = 0$ for a European call option with exercise price K.

The stochastic process $V_t = u(T - t, X_t)$ is the value of your self-financing portfolio at time $t \in [0, T]$.

The self-financing strategy (a_t, b_t) is given by

$$a_t = u_2(T - t, X_t) \quad \text{and} \quad b_t = \frac{u(T - t, X_t) - a_t X_t}{\beta_t}, \tag{4.14}$$

see (4.10) and (4.8).

At time of maturity T, the formula (4.13) yields the net portfolio value of $(X_T - K)^+$. Moreover, one can show that $a_t > 0$ for all $t \in [0, T]$, but $b_t < 0$ is not excluded. Thus short sales of stock do not occur, but borrowing money at the bond's constant interest rate $r > 0$ may become necessary.

Equation (4.13) is the celebrated *Black–Scholes option pricing formula*. We see that it is independent of the mean rate of return c for the price X_t, but it depends on the volatility σ.

If we want to understand $q = u(T, X_0)$ as a rational value in terms of *arbitrage*, suppose that the initial option price $p \neq q$. If $p > q$, apply the following strategy: at time $t = 0$

- sell the option to someone else at the price p, and

- invest q in stock and bond according to the self-financing strategy (4.14).

Thus you gain an initial net profit of $p - q > 0$. At time of maturity T, the portfolio has value $a_T X_T + b_T \beta_T = (X_T - K)^+$, and you have the obligation to pay the value $(X_T - K)^+$ to the purchaser of the option. This means: if $X_T > K$, you must buy the stock for X_T, and sell it to the option holder at the exercise price K, for a net loss of $X_T - K$. If $X_T \leq K$, you do not have to pay anything, since the option will not be exercised. Thus the total terminal profit is zero, and the net profit is $p - q$.

The scale of this game can be increased arbitrarily, by selling n options for np at time zero and by investing nq in stock and bond according to the self-financing strategy (na_t, nb_t). The net profit will be $n(p - q)$. Thus the opportunity for arbitrarily large profits exists without an accompanying risk of loss. This means arbitrage. Similar arguments apply if $q > p$; now the purchaser of the option will make arbitrarily big net profits without accompanying risks.

Notes and Comments

The idea of using Brownian motion in finance goes back to Bachelier (1900), but only after 1973, when Black, Scholes and Merton published their papers, did the theory reach a more advanced level. Since then, options, futures and many other financial derivatives have conquered the international world of finance. This led to a new, applied, dimension of an advanced mathematical theory: stochastic calculus. As we have learnt in this book, this theory requires some non-trivial mathematical tools.

In 1997, Merton and Scholes were awarded the Nobel prize for economics.

Most books, which are devoted to mathematical finance, require the knowledge of measure theory and functional analysis. For this reason they can be read only after several years of university education! Here are a few references: Duffie (1996), Musiela and Rutkowski (1997) and Karatzas and Shreve (1998), and also the chapter about finance in Karatzas and Shreve (1988).

By now, there also exist a few texts on mathematical finance which address an elementary or intermediate level. Baxter and Rennie (1996) is an easy introduction with a minimum of mathematics, but still precise and with a good explanation of the economic background. The book by Willmot, Howison and Dewynne (1995) focuses on partial differential equations and avoids stochastic calculus whenever possible. A course on finance and stochastic calculus is given by Lamberton and Lapeyre (1996) who address an intermediate level based on some knowledge of measure theory. Pliska (1997) is an introduction to finance using only discrete-time models.

4.2   A Useful Technique: Change of Measure

In this section we consider a very powerful technique of stochastic calculus: the *change of the underlying probability measure.* In the literature it often appears under the synonym *Girsanov's theorem* or *Cameron–Martin formula.*

In what follows, we cannot completely avoid measure-theoretic arguments. If you do not have the necessary background on measure theory, you should at least try to understand the main idea by considering the applications in Section 4.2.2.

4.2.1   What is a Change of the Underlying Measure?

> The main idea of the change of measure technique consists of introducing a new probability measure via a so-called *density function* which is in general not a probability density function.

We start with a simple example of two distributions on the real line. Recall the probability density of a normal $N(\mu, \sigma^2)$ random variable:

$$\varphi_{\mu,\sigma^2}(x) = \frac{1}{\sqrt{2\pi}\sigma} \exp\left\{-\frac{(x-\mu)^2}{2\sigma^2}\right\}, \quad x \in \mathbb{R},$$

and write

$$\Phi_{\mu,\sigma^2}(x) = \int_{-\infty}^{x} \varphi_{\mu,\sigma^2}(y)\, dy, \quad x \in \mathbb{R},$$

for the corresponding distribution function. Consider two pairs (μ_1, σ_1^2) and (μ_2, σ_2^2) of parameters and define

$$f_1(x) = \frac{\varphi_{\mu_1,\sigma_1^2}(x)}{\varphi_{\mu_2,\sigma_2^2}(x)} \quad \text{and} \quad f_2(x) = \frac{\varphi_{\mu_2,\sigma_2^2}(x)}{\varphi_{\mu_1,\sigma_1^2}(x)}.$$

Obviously,

$$\Phi_{\mu_1,\sigma_1^2}(x) = \int_{-\infty}^{x} f_1(y)\, \varphi_{\mu_2,\sigma_2^2}(y)\, dy, \tag{4.15}$$

and

$$\Phi_{\mu_2,\sigma_2^2}(x) = \int_{-\infty}^{x} f_2(y)\, \varphi_{\mu_1,\sigma_1^2}(y)\, dy. \tag{4.16}$$

The function f_1 is called the density function of Φ_{μ_1,σ_1^2} with respect to Φ_{μ_2,σ_2^2}, and f_2 is the density function of Φ_{μ_2,σ_2^2} with respect to Φ_{μ_1,σ_1^2}. Clearly, f_1 and f_2 are not probability densities; they are positive functions, but the integrals $\int_{-\infty}^{\infty} f_i(x)\, dx$, $i = 1, 2$, are not equal to 1.

Now we consider a more general context:

Let P and Q be two probability measures on the σ-field $\mathcal{F}$. If there exists a non-negative function f_1 such that

$$Q(A) = \int_A f_1(\omega)\, dP(\omega), \quad A \in \mathcal{F}, \tag{4.17}$$

we say that f_1 *is the density of Q with respect to P* and we also say that Q *is absolutely continuous with respect to P.*

The integrals in (4.17) have to be interpreted in the measure-theoretic sense.

In a similar way, changing the rôles of P and Q, we can introduce the *density f_2 of P with respect to Q*, given such a non-negative function exists.

If P is absolutely continuous with respect to Q, and Q is absolutely continuous with respect to P, we say that P *and Q are equivalent probability measures.*

From (4.15) and (4.16) we may conclude that two Gaussian probability measures on the real line are equivalent.

For a characterization of absolute continuity via the Radon–Nikodym theorem, see Appendix A5.

As usual, $B = (B_t, t \in [0, T])$ denotes standard Brownian motion. The definition of Brownian motion on p. 33 heavily depends on the underlying probability measure P. Indeed, for the definition of the independent, stationary increments of B and for its Gaussian distribution we must know the probability measure P on the σ-field $\mathcal{F}$. Usually, we do not pay attention to this fact, P *is simply given*, but in what follows, it will be a crucial aspect. For example, consider the one-dimensional distribution function $P(B_t \leq x)$, $x \in \mathbb{R}$. It is the distribution function of an $N(0, t)$ random variable, but if we were to change P for another probability measure Q, this function could differ from a normal distribution function. At the moment, this remark may sound a little abstract, but we will see soon that the change of measure is a very useful technique.

We are interested in processes of the form

$$\widetilde{B}_t = B_t + q\,t, \quad t \in [0, T], \tag{4.18}$$

for some constant q. With the only exception when $q = 0$, $\widetilde{B}$ is *not* standard Brownian motion. However, if we change the underlying probability measure P for an appropriate probability measure Q, $\widetilde{B}$ can be shown to be a standard Brownian motion *under the new probability measure Q*. This simply means that $\widetilde{B}$ satisfies the defining properties of a standard Brownian motion on p. 33, when you replace P with Q. This is the content of the following famous result; see for example Karatzas and Shreve (1988) for a proof. As usual,

$$\mathcal{F}_t = \sigma(B_s, s \leq t), \quad t \in [0, T], \tag{4.19}$$

is the Brownian filtration.

Girsanov's Theorem:

The following statements hold:

- The stochastic process

$$M_t = \exp\left\{-q\,B_t - \frac{1}{2}\,q^2\,t\right\}, \quad t \in [0, T], \tag{4.20}$$

  is a martingale with respect to the natural Brownian filtration (4.19) under the probability measure P.

- The relation

$$Q(A) = \int_A M_T(\omega)\, dP(\omega)\,, \quad A \in \mathcal{F}\,, \qquad (4.21)$$

  defines a probability measure Q on $\mathcal{F}$ which is equivalent to P.

- Under the probability measure Q, the process $\widetilde{B}$ defined by (4.18) is a standard Brownian motion.

- The process $\widetilde{B}$ is adapted to the filtration (4.19).

The probability measure Q is called an *equivalent martingale measure*.

The change of measure serves the purpose of eliminating the drift term in a stochastic differential equation. This will be seen in the applications below. We illustrate this fact by the following simple example.

Example 4.2.1 (Elimination of the drift in a linear stochastic differential equation)

Consider the linear stochastic differential equation

$$dX_t = cX_t\, dt + \sigma X_t\, dB_t\,, \quad t \in [0, T]\,, \qquad (4.22)$$

with constant coefficients c and $\sigma > 0$. We know from Example 3.2.4 that this equation has solution

$$X_t = X_0\, e^{(c - 0.5\sigma^2)t + \sigma B_t}\,, \quad t \in [0, T]\,. \qquad (4.23)$$

Introduce

$$\widetilde{B}_t = B_t + (c/\sigma)\, t\,, \quad t \in [0, T]\,,$$

and rewrite (4.22) as follows:

$$dX_t = \sigma X_t\, d[(c/\sigma)\, t + \sigma\, B_t] = \sigma X_t\, d\widetilde{B}_t\,, \quad t \in [0, T]\,. \qquad (4.24)$$

By Girsanov's theorem, $\widetilde{B}$ is a standard Brownian motion under the equivalent martingale measure given by (4.21) with $q = c/\sigma$. The linear stochastic differential equation (4.24) does not have a drift term; its solution X is a martingale under Q, but it is not a martingale under P. You can read off the solution X of (4.24) from (4.23) (choose $c = 0$ and replace B with $\widetilde{B}$):

$$\begin{aligned}
X_t &= X_0\, e^{-0.5\sigma^2 t + \sigma \widetilde{B}_t} \\
&= X_0\, e^{(c - 0.5\sigma^2)t + \sigma B_t}\,, \quad t \in [0, T]\,.
\end{aligned}$$

This is the solution to the original stochastic differential equation (4.22) driven by B.

It seems that we did not gain much. However, if we had known the solution (4.23) only for $c = 0$, we could have derived the solution of (4.23) for general c from the solution in the case $c = 0$. Moreover, since X is a martingale under Q, one can make use of the martingale property for proving various results about X. We will use this trick for pricing a European call option in Section 4.2.2.

□

4.2.2 An Interpretation of the Black–Scholes Formula by Change of Measure

In this section we review the Black–Scholes option pricing formula. We will show that it can be interpreted as the conditional expectation of the discounted overshoot $(X_T - K)^+$ at maturity.

First recall:

The Black–Scholes Model

- The price of one share of the risky asset (stock) is described by the stochastic differential equation

$$dX_t = cX_t \, dt + \sigma X_t \, dB_t, \quad t \in [0, T], \tag{4.25}$$

  where c is the *mean rate of return*, σ the *volatility*, B is standard Brownian motion (under P) and T is the *time of maturity* of the option.

- The price of the riskless asset (bond) is described by the deterministic differential equation

$$d\beta_t = r\beta_t \, dt, \quad t \in [0, T],$$

  where $r > 0$ is the *interest rate* of the bond.

- Your *portfolio* at time t consists of a_t shares of stock and b_t shares of bond. Thus its value at time t is given by

$$V_t = a_t X_t + b_t \beta_t, \quad t \in [0, T].$$

- The portfolio is *self-financing*, i.e.

$$dV_t = a_t \, dX_t + b_t \, d\beta_t, \quad t \in [0, T].$$

- At time of maturity, V_T is equal to the *contingent claim* $h(X_T)$ for a given function h. For a European call option, $h(x) = (x - K)^+$, where K is the *strike price* of the option, and for a European put option, $h(x) = (K - x)^+$.

In Section 4.1.4 we derived a rational price for a European call option and showed that there exists a self-financing strategy (a_t, b_t) such that $V_T = (X_T - K)^+$. In what follows, we give an intuitive interpretation of this pricing formula by applying the Girsanov theorem.

If we were naive we could argue as follows. Your gain from the option at time of maturity is $(X_T - K)^+$. In order to determine the value of this amount of money at time zero, you have to discount it with given interest rate r:

$$e^{-rT}(X_T - K)^+ . \tag{4.26}$$

We do not know X_T in advance, so let's assume that X satisfies the linear stochastic differential equation (4.25), and simply take the expectation of (4.26) as the price for the option at time zero.

This sounds convincing, although we did not use any theory. And because we did not apply any theory we will see that we are slightly wrong with our arguments, i.e. we do not get the Black–Scholes price in this way. It will turn out that the rational Black–Scholes price is the expected value of (4.26), but we will have to adjust our expectation by changing the underlying probability measure; see (4.29) for the correct formula.

This change of the probability measure P will be provided in such a way that the *discounted price* of one share of stock

$$\tilde{X}_t = e^{-rt} X_t , \quad t \in [0, T] ,$$

will become a martingale under the new probability measure Q. Write

$$f(t, x) = e^{-rt} x$$

and apply the Itô lemma (2.31) on p. 120 (notice that $f_{22}(t, x) = 0$) to obtain

$$
\begin{aligned}
d\tilde{X}_t &= -re^{-rt} X_t \, dt + e^{-rt} \, dX_t & (4.27) \\
&= -re^{-rt} X_t \, dt + e^{-rt} X_t \left[c \, dt + \sigma \, dB_t \right] \\
&= \tilde{X}_t \left[(c - r) \, dt + \sigma \, dB_t \right] \\
&=: \sigma \tilde{X}_t \, d\tilde{B}_t , & (4.28)
\end{aligned}
$$

where
$$\widetilde{B}_t = B_t + [(c - r)/\sigma]\, t\,, \quad t \in [0, T]\,.$$

From Girsanov's theorem we know that there exists an equivalent martingale measure Q which turns $\widetilde{B}$ into standard Brownian motion. The solution of (4.28), given by
$$\widetilde{X}_t = \widetilde{X}_0\, e^{-0.5\sigma^2 t + \sigma \widetilde{B}_t}\,, \quad t \in [0, T]\,,$$

turns under Q into a martingale with respect to the natural Brownian filtration.

The existence of the equivalent martingale measure allows for an intuitive interpretation of the Black–Scholes formula:

Assume in the Black–Scholes model that there exists a self-financing strategy (a_t, b_t) such that the value V_t of your portfolio at time t is given by
$$V_t = a_t X_t + b_t \beta_t\,, \quad t \in [0, T]\,,$$

and that V_T is equal to the contingent claim $h(X_T)$.

Then the value of the portfolio at time t is given by
$$V_t = E_Q\left[e^{-r(T-t)}\, h(X_T)\,|\, \mathcal{F}_t\right]\,, \quad t \in [0, T]\,, \qquad (4.29)$$

where $E_Q(A\,|\,\mathcal{F}_t)$ denotes the conditional expectation of the random variable A, given $\mathcal{F}_t = \sigma(B_s,\, s \leq t)$, under the new probability measure Q.

Until now, the conditional expectation $E(A\,|\,\mathcal{F}_t) = E_P(A\,|\,\mathcal{F}_t)$ was always defined under the original probability measure P, and so we did not indicate this dependence in our notation.

First we want to show that (4.29) is correct, and then we will use (4.29) to evaluate the price of a European call option.

Consider the discounted value process
$$\widetilde{V}_t = e^{-rt}\, V_t = e^{-rt}\,(a_t X_t + b_t \beta_t)\,.$$

The Itô lemma (2.31) yields
$$d\widetilde{V}_t = -r\widetilde{V}_t\, dt + e^{-rt}\, dV_t\,.$$

Now we use the fact that (a_t, b_t) is self-financing together with (4.27):
$$\begin{aligned}
d\widetilde{V}_t &= -re^{-rt}\,(a_t X_t + b_t \beta_t)\, dt + e^{-rt}\,(a_t\, dX_t + b_t\, d\beta_t) \\
&= a_t\,(-re^{-rt} X_t\, dt + e^{-rt}\, dX_t) \\
&= a_t\, d\widetilde{X}_t\,. \qquad (4.30)
\end{aligned}$$

Notice that $\tilde{V}_0 = V_0$. From (4.30) and (4.28) we have

$$\tilde{V}_t = V_0 + \int_0^t a_s \, d\tilde{X}_s$$

$$= V_0 + \sigma \int_0^t a_s \tilde{X}_s \, d\tilde{B}_s. \qquad (4.31)$$

Under the equivalent martingale measure Q, $\tilde{B}$ is standard Brownian motion and the process $(a_t \tilde{X}_t, t \in [0, T])$ is adapted to $(\mathcal{F}_t, t \in [0, T])$. Hence (4.31) constitutes a martingale with respect to $(\mathcal{F}_t)$. This is one of the basic properties of the Itô integral; see p. 111. In particular, the martingale property implies that

$$\tilde{V}_t = E_Q(\tilde{V}_T \mid \mathcal{F}_t), \quad t \in [0, T],$$

but

$$\tilde{V}_T = e^{-rT} V_T = e^{-rT} h(X_T).$$

Hence

$$e^{-rt} V_t = E_Q \left[e^{-rT} h(X_T) \mid \mathcal{F}_t \right],$$

or, equivalently, the value of the portfolio is given by (4.29).

Now we want to calculate the value V_t of our portfolio and the Black–Scholes price in the case of a European option.

Example 4.2.2 (The value of a European option)
Write

$$\theta = T - t \quad \text{for} \quad t \in [0, T].$$

By (4.29), the value V_t of the portfolio at time t corresponding to the contingent claim $V_T = h(X_T)$ is given by

$$V_t = E_Q \left[e^{-r\theta} h(X_T) \mid \mathcal{F}_t \right]$$

$$= E_Q \left[e^{-r\theta} h \left(X_t \, e^{(r - 0.5\sigma^2) \theta + \sigma(\tilde{B}_T - \tilde{B}_t)} \right) \mid \mathcal{F}_t \right].$$

At time t, X_t is a function of B_t, hence $\sigma(X_t) \subset \mathcal{F}_t$, and so we can treat it under $\mathcal{F}_t$ as if it was a constant. Moreover, under Q, $\tilde{B}_T - \tilde{B}_t$ is independent of $\mathcal{F}_t$ and has an $N(0, \theta)$ distribution. An application of Rule 7 on p. 72 yields that

$$V_t = f(t, X_t),$$

where

$$f(t, x) = e^{-r\theta} \int_{-\infty}^{\infty} h \left(x \, e^{(r - 0.5\sigma^2)\theta + \sigma y \theta^{1/2}} \right) \varphi(y) dy,$$

and $\varphi(y)$ denotes the standard normal density. For a European call option,

$$h(x) = (x - K)^+ = \max(0, x - K),$$

and so

$$
\begin{aligned}
f(t, x) &= \int_{-z_2}^{\infty} \left[x\, e^{-0.5\sigma^2\theta + \sigma\, y\, \theta^{1/2}} - K e^{-r\theta} \right] \varphi(y) dy \\
&= x\, \Phi(z_1) - K e^{-r\theta} \Phi(z_2),
\end{aligned}
$$

where $\Phi(x)$ is the standard normal distribution function,

$$z_1 = \frac{\ln(x/K) + (r + 0.5\sigma^2)\theta}{\sigma\theta^{1/2}} \quad \text{and} \quad z_2 = z_1 - \sigma\theta^{1/2}.$$

This is exactly the formula (for $t = 0$) that we derived on p. 174 for the rational price of a European call option.

The price of the European put option (with $h(x) = (K - x)^+$) can be calculated in the same way with

$$f(t, x) = K e^{-r\theta} \Phi(-z_2) - x\, \Phi(-z_1).$$

Verify this formula!   □

Notes and Comments

The derivation of the Black–Scholes price via a change of the underlying measure (also called *change of numeraire*) was initiated by the fundamental paper by Harrison and Pliska (1981). Since then this technique belongs to the standard repertoire of mathematical finance; see for example Karatzas and Shreve (1998), Lamberton and Lapeyre (1996) or Musiela and Rutkowski (1997). The interpretation of the Black–Scholes formula as a conditional expectation and related results can be found in any of these books.

A proof of Girsanov's theorem can be found in advanced textbooks on probability theory and stochastic processes; see for example Kallenberg (1997) or Karatzas and Shreve (1988).

Appendix

A1 Modes of Convergence

The following theory can be found for instance in Feller (1968), Karr (1993) or Loève (1978).

We introduce the main modes of convergence for a sequence of random variables $A, A_1, A_2, \ldots$.

Convergence in Distribution

The sequence (A_n) *converges in distribution* or *converges weakly to the random variable A* ($A_n \xrightarrow{d} A$) if for all bounded, continuous functions f the relation

$$Ef(A_n) \to Ef(A), \quad n \to \infty,$$

holds.

Notice: $A_n \xrightarrow{d} A$ holds if and only if for all continuity points x of the distribution function F_A the relation

$$F_{A_n}(x) \to F_A(x), \quad n \to \infty, \tag{A.1}$$

is satisfied. If F_A is continuous then (A.1) can even be strenghtened to uniform convergence:

$$\sup_x |F_{A_n}(x) - F_A(x)| \to 0, \quad n \to \infty.$$

It is also well known that convergence in distribution is equivalent to pointwise convergence of the corresponding characteristic functions:

$$A_n \xrightarrow{d} A \quad \text{if and only if} \quad Ee^{itA_n} \to Ee^{itA} \quad \text{for all } t.$$

Example A1.1 (Convergence in distribution of Gaussian random variables) Assume that (A_n) is a sequence of normal $N(\mu_n, \sigma_n^2)$ random variables.

First suppose that $\mu_n \to \mu$ and $\sigma_n^2 \to \sigma^2$, where μ and σ^2 are finite numbers. Then the corresponding characteristic functions converge for every $t \in \mathbb{R}$:

$$Ee^{itA_n} = e^{it\mu_n - 0.5\sigma_n^2 t^2} \to e^{it\mu - 0.5\sigma^2 t^2}.$$

The right-hand side is the characteristic function of an $N(\mu, \sigma^2)$ random variable A. Hence $A_n \xrightarrow{d} A$.

Also the converse is true. If we know that $A_n \xrightarrow{d} A$, then the characteristic functions $e^{it\mu_n - 0.5\sigma_n^2 t^2}$ necessarily converge for every t. From this fact we conclude that there exist real numbers μ and σ^2 such that $\mu_n \to \mu$ and $\sigma_n^2 \to \sigma^2$. This implies that A is necessarily a normal $N(\mu, \sigma^2)$ random variable. $\square$

Convergence in Probability

The sequence (A_n) *converges in probability to the random variable A* $(A_n \xrightarrow{P} A)$ if for all positive ε the relation

$$P(|A_n - A| > \varepsilon) \to 0, \quad n \to \infty,$$

holds.

Convergence in probability implies convergence in distribution. The converse is true if and only if $A = a$ for some constant a.

Almost Sure Convergence

The sequence (A_n) *converges almost surely (a.s.) or with probability 1 to the random variable A* $(A_n \xrightarrow{a.s.} A)$ if the set of ωs with

$$A_n(\omega) \to A(\omega), \quad n \to \infty,$$

has probability 1.

> This means that
>
> $$P(A_n \to A) = P(\{\omega : A_n(\omega) \to A(\omega)\}) = 1.$$

Convergence with probability 1 implies convergence in probability, hence convergence in distribution. Convergence in probability does not imply convergence a.s. However, $A_n \xrightarrow{P} A$ implies that $A_{n_k} \xrightarrow{a.s.} A$ for a suitable subsequence (n_k).

L^p-Convergence

> Let $p > 0$. The sequence (A_n) *converges in L^p or in pth mean to A* $(A_n \xrightarrow{L^p} A)$ if $E[|A_n|^p + |A|^p] < \infty$ for all n and
>
> $$E|A_n - A|^p \to 0, \quad n \to \infty.$$

By Markov's inequality, $P(|A_n - A| > \varepsilon) \leq \varepsilon^{-p} E|A_n - A|^p$ for positive p and ε. Thus $A_n \xrightarrow{L^p} A$ implies that $A_n \xrightarrow{P} A$. The converse is in general not true.

For $p = 2$, we say that (A_n) *converges in mean square to A*. This notion can be extended to stochastic processes; see for example Appendix A4. Mean square convergence is convergence in the Hilbert space

$$L^2 = L^2[\Omega, \mathcal{F}, P] = \{X : EX^2 < \infty\}$$

endowed with the inner product $<X, Y> = E(XY)$ and the norm $\|X\| = \sqrt{(X, X)}$. The symbol X stands for the equivalence class of random variables Y satisfying $X \stackrel{d}{=} Y$.

A2   Inequalities

In this section we give some standard inequalities which are frequently used in this book.

> **The Chebyshev inequality:**
>
> $$P(|X - EX| > x) \leq x^{-2} \operatorname{var}(X), \quad x > 0.$$

The Cauchy–Schwarz inequality:

$$E|XY| \leq (EX^2)^{1/2} (EY^2)^{1/2}.$$

The Jensen inequality:

Let f be a convex function on $\mathbb{R}$. If $E|X|$ and $E|f(X)|$ are finite, then

$$f(EX) \leq Ef(X).$$

In particular,

$$(E|X|^q)^{1/q} \leq (E|X|^p)^{1/p} \quad \text{for} \quad 0 < q < p.$$

This is *Lyapunov's inequality*.

Jensen's inequality remains valid for conditional expectations: let $\mathcal{F}$ be a σ-field on Ω. Then

$$f(E(X\,|\,\mathcal{F})) \leq E(f(X)\,|\,\mathcal{F}). \tag{A.2}$$

In particular,

$$|E(X\,|\,\mathcal{F})| \leq E(|X|\,|\,\mathcal{F}) \quad \text{and} \quad [E(X\,|\,\mathcal{F})]^2 \leq E(X^2\,|\,\mathcal{F}).$$

A3   Non-Differentiability and Unbounded Variation of Brownian Sample Paths

Let $B = (B_t, t \geq 0)$ be Brownian motion. Recall the definitions of an H-self-similar process from (1.12) on p. 36 and of a process with stationary increments from p. 30. We also know that Brownian motion is a 0.5-self-similar process with stationary, independent increments; see Section 1.3.1. We show the non-differentiability of Brownian sample paths in the more general context of self-similar processes.

Proposition A3.1 (Non-differentiability of self-similar processes)
Suppose (X_t) is H-self-similar with stationary increments for some $H \in (0, 1)$. Then for every fixed t_0,

$$\limsup_{t \downarrow t_0} \frac{|X_t - X_{t_0}|}{t - t_0} = \infty,$$

i.e. sample paths of H-self-similar processes are nowhere differentiable with probability 1.

Proof. Without loss of generality we choose $t_0 = 0$. Let (t_n) be a sequence such that $t_n \downarrow 0$. Then, by H-self-similarity, $X_0 = 0$ a.s., and hence

$$P\left(\lim_{n \to \infty} \sup_{0 \le s \le t_n} \left|\frac{X_s}{s}\right| > x\right) = \lim_{n \to \infty} P\left(\sup_{0 \le s \le t_n} \left|\frac{X_s}{s}\right| > x\right)$$

$$\ge \limsup_{n \to \infty} P\left(\left|\frac{X_{t_n}}{t_n}\right| > x\right)$$

$$= \limsup_{n \to \infty} P\left(t_n^{H-1}|X_1| > x\right)$$

$$= 1, \quad x > 0.$$

Hence, with probability 1, $\limsup_{n \to \infty} |X_{t_n}/t_n| = \infty$ for any sequence $t_n \downarrow 0$. $\square$

Proposition A3.2 (Unbounded variation of Brownian sample paths)
For almost all Brownian sample paths,

$$v(B(\omega)) = \sup_{\tau} \sum_{i=1}^{n} \left|B_{t_i}(\omega) - B_{t_{i-1}}(\omega)\right| = \infty \quad \text{a.s.},$$

where the supremum is taken over all possible partitions $\tau : 0 = t_0 < \cdots < t_n = T$ of $[0, T]$.

Proof. For convenience, assume $T = 1$. Suppose that $v(B(\omega)) < \infty$ for a given ω. Let (τ_n) be a sequence of partitions $\tau_n : 0 = t_0 < t_1 < \cdots < t_{n-1} < t_n = 1$, such that $\text{mesh}(\tau_n) \to 0$. Recall that $\Delta_i B = B_{t_i} - B_{t_{i-1}}$. The following chain of inequalities holds:

$$Q_n(\omega) = \sum_{i=1}^{n} (\Delta_i B(\omega))^2$$

$$\le \max_{i=1,\ldots,n} |\Delta_i B(\omega)| \sum_{i=1}^{n} |\Delta_i B(\omega)|$$

$$\le \max_{i=1,\ldots,n} |\Delta_i B(\omega)| \, v(B(\omega)). \tag{A.3}$$

Since B has continuous sample paths with probability 1, we may assume that $B_t(\omega)$ is a continuous function of t. It is also uniformly continuous on $[0, 1]$ which, in combination with $\text{mesh}(\tau_n) \to 0$, implies that $\max_{i=1,\ldots,n} |\Delta_i B(\omega)| \to 0$. Hence the right-hand side of (A.3) converges to zero, implying that

$$Q_n(\omega) \to 0. \tag{A.4}$$

On the other hand, we know from p. 98 that $Q_n \xrightarrow{P} 1$, hence $Q_{n_k}(\omega) \xrightarrow{a.s.} 1$ for a suitable subsequence (n_k); see p. 187. Thus (A.4) is only possible on a null-set, and so

$$P(\{\omega : v(B(\omega)) = \infty\}) = 1.$$ □

A4   Proof of the Existence of the General Itô Stochastic Integral

In this section we give a proof of the existence of the Itô stochastic integral. This requires some knowledge of the spaces L^2 of square integrable functions and of measure-theoretic arguments. If you think this is too much for you, you should avoid this section, but a short glance at the proof would be useful. We do not give all the details; sometimes we refer to some other books.

As usual, $B = (B_t, t \geq 0)$ is Brownian motion, and $(\mathcal{F}_t, t \geq 0)$ is the corresponding natural filtration. Let $C = (C_t, t \in [0, T])$ be a stochastic process satisfying the Assumptions on p. 108, which we recall here for convenience:

- C_t is a function of B_s, $s \leq t$.

- $\int_0^T EC_s^2 < \infty$.

All stochastic processes are supposed to live on the same probability space $[\Omega, \mathcal{F}, P]$ (P is a probability measure on the σ-field $\mathcal{F}$).

Lemma A4.1 *If C satisfies the Assumptions, then there exists a sequence $(C^{(n)})$ of simple processes (see p. 101 for the definition) such that*

$$\int_0^T E[C_s - C_s^{(n)}]^2 \, ds \to 0. \tag{A.5}$$

A proof can be found, for example, in Kloeden and Platen (1992), Lemma 3.2.1.

Thus simple processes $C^{(n)}$ are dense in the space $L^2[\Omega \times [0, T], dP \times dt]$. Relation (A.5) means exactly this: in the norm of this space, C can be approximated arbitrarily well by simple processes $C^{(n)}$.

As usual, denote by

$$I_t(C^{(n)}) = \int_0^t C_s^{(n)} \, dB_s, \quad t \in [0, T],$$

the Itô stochastic integral of the simple process $C^{(n)}$. It is well defined as a Riemann–Stieltjes sum; cf. (2.11).

Lemma A4.2 *Under the Assumptions, the following relation holds for all* $n, k \in \mathbb{N}$:

$$E \sup_{0 \leq t \leq T} [I_t(C^{(n+k)}) - I_t(C^{(n)})]^2 \leq 4 E \int_0^T [C_t^{(n+k)} - C_t^{(n)}]^2 \, dt. \quad (A.6)$$

Proof. Both, $I(C^{(n)})$ and $I(C^{(n+k)})$ are martingales with respect to Brownian motion, hence $I(C^{(n+k)}) - I(C^{(n)}) = I(C^{(n+k)} - C^{(n)})$ is a martingale with respect to $(\mathcal{F}_t)$.

For a general martingale $(M, (\mathcal{F}_t))$ on $[0, T]$, Doob's squared maximal inequality (see for example Dellacherie and Meyer (1982), Theorem 99 on p. 164) tells us that

$$E \sup_{0 \leq t \leq T} M_t^2 \leq 4 E M_T^2.$$

In particular,

$$E \sup_{0 \leq t \leq T} [I_t(C^{(n+k)} - C^{(n)})]^2 \leq 4 E [I_T(C^{(n+k)} - C^{(n)})]^2. \quad (A.7)$$

By the isometry property (2.14), applied to the simple process $I(C^{(n+k)} - C^{(n)})$, the right-hand side becomes

$$4 E \int_0^T [C_t^{(n+k)} - C_t^{(n)}]^2 \, dt,$$

which, together with (A.7), proves (A.6). $\qquad \square$

Since (A.5) holds, $(C^{(n)})$ is a Cauchy sequence in $L^2[\Omega \times [0, T], dP \times dt]$. This implies, in particular, that the right-hand side in (A.6) converges to zero. Thus we can find a subsequence (k_n), say, such that

$$E \sup_{0 \leq t \leq T} [I_t(C^{(k_{n+1})} - C^{(k_n)})]^2 \leq \frac{1}{2^n}.$$

This and the Chebyshev inequality on p. 187 imply that, for every $\varepsilon > 0$,

$$\sum_{n=1}^{\infty} P \left(\sup_{0 \leq t \leq T} |I_t(C^{(k_{n+1})}) - I_t(C^{(k_n)})| > \varepsilon \right)$$

$$\leq \frac{1}{\varepsilon^2} \sum_{n=1}^{\infty} E \sup_{0 \leq t \leq T} [I_t(C^{(k_{n+1})} - C^{(k_n)})]^2 < \infty.$$

An appeal to the Borel–Cantelli lemma (see for example Karr (1993)) yields that the processes $I(C^{(k_n)})$ converge uniformly on $[0, T]$ to a stochastic process $I(C)$, say, with probability 1. Since the (uniformly) converging processes have continuous sample paths, so has the limiting process $I(C)$.

Now, letting k in (A.6) go to infinity, we conclude that

$$E \sup_{0 \leq t \leq T} [I_t(C) - I_t(C^{(n)})]^2 \leq 4 E \int_0^T [C_t - C_t^{(n)}]^2 dt .$$

By Lemma A4.1, the right-hand side converges to zero as $n \to \infty$, and so we have proved that the limit $I(C)$ does not depend on the particular choice of the approximating sequence $(C^{(n)})$ of simple processes.

> The limit process $I(C)$ is the desired Itô stochastic integral process, denoted by
> $$I_t(C) = \int_0^t C_s \, dB_s , \quad t \in [0, T] .$$
> Its sample paths are continuous with probability 1.

Lemma A4.3 *The pair $(I(C), (\mathcal{F}_t))$ constitutes a martingale.*

Proof. We know that $(I(C^{(n)}), (\mathcal{F}_t))$ is a martingale (see p. 104) for all n. In particular, it is adapted to $(\mathcal{F}_t)$. This does not change in the limit. Moreover, for all n,

$$E(I_t(C^{(n)}) \mid \mathcal{F}_s) = I_s(C^{(n)}) \quad \text{for} \quad s < t .$$

Using Lebesgue dominated convergence and, if necessary, passing to a subsequence (k_n) such that $I_t(C^{(k_n)})$ converges to $I_t(C)$ uniformly for t, with probability 1, we obtain

$$\lim_{n \to \infty} E(I_t(C^{(k_n)}) \mid \mathcal{F}_s) = E(\lim_{n \to \infty} I_t(C^{(k_n)}) \mid \mathcal{F}_s) = E(I_t(C) \mid \mathcal{F}_s)$$

$$= \lim_{n \to \infty} I_s(C^{(k_n)}) = I_s(C) .$$

Hence $I(C)$ is a martingale.   □

Lemma A4.4 *The Itô stochastic integral $I(C)$ satisfies the isometry property*

$$E \left(\int_0^t C_s \, dB_s \right)^2 = \int_0^t E C_s^2 \, ds , \quad t \in [0, T] . \tag{A.8}$$

Proof. We know that $I(C^{(n)})$ satisfies this property; see (2.14). By (A.6), (A.5) and since the L^2-norm is continuous,

$$E[I_t(C^{(n)})]^2 \quad = E[I_t(C^{(n)} - C) + I_t(C)]^2 \quad = E[I_t(C)]^2 + o(1),$$

$$\int_0^t E[C_s^{(n)}]^2 \, ds \quad = \int_0^t E[(C_s^{(n)} - C_s) + C_s]^2 \, ds \quad = \int_0^t EC_s^2 \, ds + o(1).$$

Since the two left-hand sides coincide, by the isometry property for $I(C^{(n)})$, we obtain as $n \to \infty$ the desired isometry relation (A.8). $\qquad\Box$

Lemma A4.5 *For a simple process C, the particular Riemann–Stieltjes sums in the definition (2.11) coincide with the Itô stochastic integral $I(C)$.*

The proof makes use of the fact that simple processes $C^{(n)}$ are dense in the underlying L^2-space.

A5 The Radon–Nikodym Theorem

Let $[\Omega, \mathcal{F}']$ be a measurable space, i.e. $\mathcal{F}'$ is a σ-field on Ω. Consider two measures μ and ν on $\mathcal{F}'$. We say that μ *is absolutely continuous with respect to ν ($\mu \ll \nu$)* if

$$\text{for all } A \in \mathcal{F}': \quad \nu(A) = 0 \text{ implies that } \mu(A) = 0.$$

We say that μ and ν are *equivalent measures* if $\mu \ll \nu$ and $\nu \ll \mu$.

The Radon–Nikodym Theorem:

Assume μ and ν are two σ-finite measures. (This is in particular satisfied if they are probability measures.) Then $\mu \ll \nu$ holds if and only if there exists a non-negative measurable function f such that

$$\mu(A) = \int_A f(\omega) \, d\nu(\omega), \quad A \in \mathcal{F}'.$$

If this relation holds with f replaced by another non-negative function g, then

$$\nu(f \neq g) = 0.$$

The ν-almost everywhere unique function f is called *the density of μ with respect to ν.*

For a proof of this result, see Billingsley (1995) or Dudley (1989), Section 5.5.

A6   Proof of the Existence and Uniqueness of the Conditional Expectation

Let $[\Omega, \mathcal{F}, P]$ be a probability space, i.e. $\mathcal{F}$ is a σ-field on Ω and P is a probability measure on $\mathcal{F}$. Consider the σ-field $\mathcal{F}' \subset \mathcal{F}$ and denote by P' the restriction of P to $\mathcal{F}'$. Let X be a random variable on Ω.

Theorem A6.1 *If $E|X| < \infty$, then there exists a random variable Z such that*

(a) $\sigma(Z) \subset \mathcal{F}'$ *and*

(b) *for all $A \in \mathcal{F}'$,*

$$\int_A Z(\omega)\, dP(\omega) = \int_A X(\omega)\, dP, \quad A \in \mathcal{F}'.$$

If (a) and (b) hold with Z replaced by another random variable Z' then

$$P'(Z \neq Z') = 0. \tag{A.9}$$

We call Z the conditional expectation of X given the σ-field $\mathcal{F}'$, and we write $Z = E(X\,|\,\mathcal{F})$.

Proof. A) Assume that $X \geq 0$ P-a.s. Consider the (smaller) probability space $[\Omega, \mathcal{F}', P']$ and define the measure

$$\nu(A) = \int_A X(\omega)\, dP(\omega), \quad A \in \mathcal{F}'.$$

If $P(A) = 0$, then $X I_A = 0$ P-a.s., and so $\nu(A) = E(X I_A) = 0$. Hence $\nu \ll P'$, where both measures, P' and ν, are defined on $\mathcal{F}'$. By virtue of the Radon–Nikodym theorem on p. 193, there exists a non-negative random variable Z on $[\Omega, \mathcal{F}']$ such that

$$\nu(A) = \int_A Z(\omega)\, dP'(\omega), \quad A \in \mathcal{F}',$$

and the random variable Z is unique in the sense of (A.9). Moreover, recalling the definition of the integral and using that $\sigma(Z) \subset \mathcal{F}'$, one can show that $\int_A Z(\omega)\, dP'(\omega) = \int_A Z(\omega)\, dP(\omega)$.

B) Assume that $E|X| < \infty$. Write $X = X^+ - X^-$, where $X^+ = \max(0, X)$ and $X^- = -\min(0, X)$. Apply the first part of the theorem to X^+ and X^- separately. It yields two non-negative random variables Z^+ and Z^- such that (a) and (b) hold if we replace Z by them. Moreover, $EX^{\pm} = EZ^{\pm} < \infty$ since $E|X| = EX^+ + EX^- < \infty$. Hence $Z^{\pm} < \infty$ P-a.s., and so we can define $Z = Z^+ - Z^-$ which is the desired random variable. $\quad\square$

Bibliography

[1] Arnold, L. (1973) *Stochastische Differentialgleichungen. Theorie und Anwendungen.* R. Oldenbourg Verlag, München, Wien. *[129]*

[2] Ash, R.B. and Gardner, M.F. (1975) *Topics in Stochastic Processes.* Academic Press, New York. *[33]*

[3] Bachelier, L. (1900) Théorie de la spéculation. *Ann. Sci. École Norm. Sup.* III–17, 21–86. Translated in: Cootner, P.H. (Ed.) (1964) *The Random Character of Stock Market Prices,* pp. 17–78. MIT Press, Cambridge, Mass. *[33,42,175]*

[4] Baxter, M. and Rennie, A. (1996) *Financial Calculus. An Introduction to Derivative Pricing.* Cambridge University Press, Cambridge. *[176]*

[5] Billingsley, P. (1968) *Convergence of Probability Measures.* Wiley, New York. *[56]*

[6] Billingsley, P. (1995) *Probability and Measure,* 3rd edition. Wiley, New York. *[77,193]*

[7] Black, F. and Scholes, M. (1973) The pricing of options and corporate liabilities. *J. Political Economy* **81**, 635–654. *[1,42,167,175]*

[8] Borodin, A.N. and Salminen, P. (1996) *Handbook of Brownian Motion – Facts and Formulae.* Birkhäuser, Basel. *[52]*

[9] Chung, K.L. and Williams, R.J. (1990) *Introduction to Stochastic Integration,* 2nd edition. Birkhäuser, Boston. *[112,122,136,137]*

[10] Ciesielski, Z. (1965) *Lectures on Brownian Motion. Heat Conduction and Potential Theory.* Aarhus University Lecture Notes Series, Aarhus. *[56]*

[11] Dellacherie, C. and Meyer, P.-A. (1982) *Probabilities and Potential B.* North–Holland, Amsterdam. *[191]*

[12] Doob, J.L. (1953) *Stochastic Processes.* Wiley, New York. *[112]*

[13] Dudley, R.M. (1989) *Real Analysis and Probability.* Wadsworth, Brooks and Cole, Pacific Grove. *[193]*

[14] Dudley, R.M. and Norvaiša, R. (1998a) Product integrals, Young integrals and p-variation. Lecture Notes in Mathematics. Springer, New York. To appear. *[94,96]*

[15] Dudley, R.M. and Norvaiša, R. (1998b) A survey on differentiability of six operators in relation to probability and statistics. Lecture Notes in Mathematics. Springer, New York. To appear. *[94,96]*

[16] Duffie, D. (1996) *Dynamic Asset Pricing Theory.* Princeton University Press. *[176]*

[17] Einstein, A. (1905) On the movement of small particles suspended in a stationary liquid demanded by the molecular-kinetic theory of heat. *Ann. Phys.* **17**. *[33]*

[18] Feller, W. (1968) *An Introduction to Probability Theory and Its Applications I,II,* 3rd edition. Wiley, New York. *[185]*

[19] Gikhman, I.I. and Skorokhod, A.V. (1975) *The Theory of Stochastic Processes.* Springer, Berlin. *[33]*

[20] Grimett, G.R. and Stirzaker, D.R. (1994) *Probability and Random Processes.* Clarendon Press, Oxford. *[33]*

[21] Gut, A. (1995) *An Intermediate Course in Probability.* Springer, Berlin. *[23]*

[22] Harrison, M.J. and Pliska, S.R. (1981) Martingales and stochastic integrals in the theory of continuous trading. *Stoch. Proc. Appl.* **11**, 215–260. *[184]*

[23] Hida, T. (1980) *Brownian Motion.* Springer, New York. *[52]*

[24] Ikeda, N. and Watanabe, S. (1989) *Stochastic Differential Equations and Diffusion Processes,* 2nd edition. North–Holland, Amsterdam. *[112,122]*

[25] Itô, K. (1942) Differential equations determining Markov processes (in Japanese). *Zenkoku Shijo Sugaku Danwakai* **244**, 1352–1400. *[112]*

[26] Itô, K. (1944) Stochastic integral. *Proc. Imp. Acad. Tokyo* **20**, 519–524. *[112]*

[27] Itô, K. (1951) On a formula concerning stochastic differentials. *Nagoya Math. J.* **3**, 55–65. *[122]*

[28] Kallenberg, O. (1997) *Foundations of Modern Probability*. Springer, New York. *[184]*

[29] Karatzas, I. and Shreve, S.E. (1988) *Brownian Motion and Stochastic Calculus*. Springer, Berlin. *[56,85,112,122,176,178,184]*

[30] Karatzas, I. and Shreve, S.E. (1998) *Methods of Mathematical Finance*. Springer, Heidelberg. To appear. *[176,184]*

[31] Karlin, S. and Taylor, H.M. (1975) *A First Course in Stochastic Processes*. Academic Press, New York. *[33]*

[32] Karlin, S. and Taylor, H.M. (1981) *A Second Course in Stochastic Processes*. Academic Press, New York. *[33]*

[33] Karr, A.F. (1993) *Probability*. Springer, New York. *[185,192]*

[34] Kloeden, P.E. and Platen, E. (1992) *Numerical Solution of Stochastic Differential Equations*. Springer, Berlin. *[137,166,190]*

[35] Kloeden, P.E., Platen, E. and Schurz, H. (1994) *Numerical Solution of SDE through Computer Experiments*. Springer, Berlin. *[166]*

[36] Lamberton, D. and Lapeyre, B. (1996) *Introduction to Stochastic Calculus Applied to Finance*. Chapman and Hall, London. *[112,153,176,184]*

[37] Langevin, P. (1908) Sur la théorie du mouvement brownien. *C.R. Acad. Sci. Paris* **146**, 530–533. *[141]*

[38] Loève, M. (1978) *Probability Theory I,II*, 4th edition. Springer, Berlin. *[185]*

[39] Mendenhall, W., Wackerly, D.D. and Scheaffer, R.L. (1990) *Mathematical Statistics with Applications*, 4th edition. PWS–Kent Publishing Company, Boston. *[23]*

[40] Merton, R.C. (1973) Theory of option pricing. *Bell. J. Econ. Manag. Sci.* **4**, 141–183. *[1,42,167,175]*

[41] Musiela, M. and Rutkowski, M. (1997) *Martingale Methods in Financial Modelling*. Springer, Berlin. *[176,184]*

[42] Øksendahl, B. (1985) *Stochastic Differential Equations, an Introduction with Applications*. Springer, Berlin. *[112,122,137]*

[43] Pitman, J. (1993) *Probability*. Springer, Berlin. *[22]*

[44] Pliska, S.R. (1997) *Introduction to Mathematical Finance*. Discrete Time Models. Blackwell, Malden (MA). *[176]*

[45] Pollard, D. (1984) *Convergence of Stochastic Processes*. Springer, Berlin. *[56]*

[46] Protter, P. (1992) *Stochastic Integration and Differential Equations*. Springer, Berlin. *[96,112,122,136]*

[47] Resnick, S.I. (1992) *Adventures in Stochastic Processes*. Birkhäuser, Boston. *[33]*

[48] Revuz, D. and Yor, M. (1991) *Continuous Martingales and Brownian Motion*. Springer, Berlin. *[56,85]*

[49] Shorack, G.R. and Wellner, J.A. (1986) *Empirical Processes with Applications to Statistics*. Wiley, New York. *[41]*

[50] Stratonovich, R.L. (1966) A new representation for stochastic integrals and equations. *SIAM J. Control* **4**, 362–371. *[129]*

[51] Taylor, S.J. (1972) Exact asymptotic estimates of Brownian path variation. *Duke Math. J.* **39**, 219–241. *[95]*

[52] Wiener, N. (1923) Differential space. *J. Math. Phys.* **2**, 131–174. *[33]*

[53] Williams, D. (1991) *Probability with Martingales*. Cambridge University Press, Cambridge, UK. *[77,85]*

[54] Willmot, P., Howison, S. and Dewynne, J. (1995) *The Mathematics of Financial Derivatives*. Cambridge University Press. *[176]*

[55] Young, L.C. (1936) An inequality of the Hölder type, connected with Stieltjes integration. *Acta Math.* **67**, 251–282. *[94,96]*

[56] Zauderer, E. (1989) *Partial Differential Equations of Applied Mathematics*, 2nd edition. Wiley, Singapore. *[174]*

Index

List of Abbreviations and Symbols

I have tried as much as possible to use uniquely defined abbreviations and symbols. In various cases, however, symbols can have different meanings in different sections. The list below gives the most typical usage. Commonly used mathematical symbols are not explained.

Symbol	Explanation	p.
B, B_t	Brownian motion	33
$Bin(n, p)$	binomial distribution with parameters n and p	9
CLT	central limit theorem	45
$\text{corr}(X, Y)$	correlation of the random variables X and Y	18
$\text{cov}(X, Y)$	covariance of the random variables X and Y	18
$c_X(t, s)$	covariance function of the process X	28
Δ_i	length $t_i - t_{i-1}$ of the interval $[t_{i-1}, t_i]$	
$\Delta_i f$	increment of the function f on $[t_{i-1}, t_i]$: $\Delta_i f = f(t_i) - f(t_{i-1})$	
EX	expectation of the random variable X	12
$E\mathbf{X}$	expectation of the random vector $\mathbf{X}$	17
$E(X \,|\, Y)$	conditional expectation of the random variable X, given the random variable, the random vector or the stochastic process Y	69
$E(X \,|\, \mathcal{F})$	conditional expectation of the random variable X given the σ-field $\mathcal{F}$	68
$Exp(\lambda)$	exponential distribution with parameter λ	32
FCLT	functional central limit theorem	47
fidis	finite-dimensional distributions of a stochastic process	27
f_i, f_{ij}, f_{ijk}	partial derivatives of f with respect to the ith, i, jth, i, j, kth variables	

$Poi(\lambda)$	Poisson distribution with parameter λ	9
$\mathbb{R}$	real line	
$\mathbb{R}^n$	n-dimensional Euclidean space	
$S(C)$	Stratonovich integral of the process C	124
σ_X^2	variance of the random variable X	12
$\sigma_X^2(t)$	variance function of the process X	28
$\Sigma_{\mathbf{X}}$	covariance matrix of the random vector $\mathbf{X}$	17
$\sigma(\mathcal{C})$	σ-field generated by the collection of sets $\mathcal{C}$	63
$\sigma(Y)$	σ-field generated by a random variable, a random vector or a stochastic process Y	66
sup	$a = \sup_n a_n$: the supremum of a sequence of real numbers a_n. If $a \in \mathbb{R}$, $a \geq a_n$ for all n and for every $\varepsilon > 0$ there exists a k such that $a - \varepsilon < a_k$. If $a = \infty$, then for every $M > 0$ there exists a k such that $a_k > M$. For a finite index set I, $\sup_{n \in I} a_n = \max_{n \in I} a_n$.	
$U(a,b)$	uniform random variable on (a,b)	11
$\text{var}(X)$	variance of the random variable X	12
X	random variable or stochastic process	
$\mathbf{X}$	random vector	
$\mathbb{Z}$	set of integers	
$\approx$	$a(x) \approx b(x)$ means that $a(x)$ is approximately (roughly) of the same order as $b(x)$. It is only used in a heuristic sense.	
$[x]$	integer part of x	
$\{x\}$	fractional part of x	
x^+	$x^+ = \max(0, x)$	
x^-	$x^- = -\min(0, x)$	
$\mathbf{x}'$	transpose of the vector $\mathbf{x}$	
A^c	complement of the set A	
$\xrightarrow{\text{a.s.}}$	$A_n \xrightarrow{\text{a.s.}} A$: a.s. convergence	186
$\xrightarrow{d}$	$A_n \xrightarrow{d} A$: convergence in distribution	185
$\xrightarrow{L^p}$	$A_n \xrightarrow{L^p} A$: convergence in L^p, pth mean convergence	187
$\xrightarrow{L^2}$	$A_n \xrightarrow{L^2} A$: convergence in L^2, mean square convergence	187
$\xrightarrow{P}$	$A_n \xrightarrow{P} A$: convergence in probability	186
$\stackrel{d}{=}$	$A \stackrel{d}{=} B$: the random elements (random variables, random vectors, stochastic processes) A and B have the same distribution, i.e. $P(A \in C) = P(B \in C)$ for all suitable sets C. For random variables and random vectors A, B this means that their distribution functions are the same.	

$\#A$ the number of elements in the set A
• $C \bullet B$: martingale transform   83

CPSIA information can be obtained
at www.ICGtesting.com
Printed in the USA
LVHW081946221221
706979LV00002B/13